Riawa Smith
331-8875

PHalarope Books are designed specifically for the amateur naturalist. These volumes represent excellence in natural history publishing. Each book in the PHalarope series is based on a nature course or program at the college or adult education level or is sponsored by a museum or nature center. Each PHalarope Book reflects the author's teaching ability as well as writing ability.

Among the books in the series:

The Amateur Naturalist's Handbook
Vinson Brown

The Amateur Naturalist's Diary
Vinson Brown

Nature Drawing: A Tool for Learning
Clare Walker Leslie

Outdoor Education: A Manual for Teaching in Nature's Classroom
Michael Link, Director, Northwoods Audubon Center, Minnesota

Nature with Children of All Ages: Activities and Adventures for Exploring, Learning, & Enjoying the World Around Us
Edith A. Sisson, Massachusetts Audubon Society

The Wildlife Observer's Guidebook
Charles E. Roth, Massachusetts Audubon Society

A Complete Manual of Amateur Astronomy: Tools and Techniques for Astronomical Observations
P. Clay Sherrod with Thomas L. Koed

365 Starry Nights: An Introduction to Astronomy for Every Night of the Year
Chet Raymo

At the Sea's Edge: An Introduction to Coastal Oceanography for the Amateur Naturalist
William T. Fox

Nature in the Northwest: An Introduction to the Natural History and Ecology of the Northwestern United States from the Rockies to the Pacific
Susan Schwartz

James Reid Macdonald, former curator of paleontology at the Museum of Geology, South Dakota School of Mines and Technology, was also a professor of paleontology at the University of Idaho, as well as a senior curator at the Los Angeles County Museum of Natural History. He has also taught geology at the University of Southern California.

THE FOSSIL COLLECTOR'S HANDBOOK

A Paleontology Field Guide

James Reid Macdonald

A Spectrum Book
Prentice-Hall, Inc., Englewood Cliffs, New Jersey 07632

Library of Congress Cataloging in Publication Data

Macdonald, J. R. (James Reid) (date)
The fossil collector's handbook.

(PHalarope books)
"A Spectrum Book."
Includes index.
1. Paleontology—Collectors and collecting—United States. I. Title.
QE746.M3 1983 560'.75'0973 83-11140
ISBN 0-13-329235-5
ISBN 0-13-329227-4 (pbk.)

10 9 8 7 6 5 4 3 2 1

ISBN 0-13-329235-5

ISBN 0-13-329227-4 {PBK.}

A Spectrum Book. Printed in the United States of America.

Editorial/production supervision by Peter Jordan
Cover design by Hal Siegel
Manufacturing buyer: Pat Mahoney

This book is available at a special discount when ordered in bulk quantities. Contact Prentice-Hall, Inc., General Publishing Division, Special Sales, Englewood Cliffs, N.J. 07632.

Prentice-Hall International, Inc., London
Prentice-Hall of Australia Pty. Limited, Sydney
Prentice-Hall Canada Inc., Toronto
Prentice-Hall of India Private Limited, New Delhi
Prentice-Hall of Japan, Inc., Tokyo
Prentice-Hall of Southeast Asia Pte. Ltd., Singapore
Whitehall Books Limited, Wellington, New Zealand
Editora Prentice-Hall do Brasil Ltda., Rio de Janeiro

For Philo, Reid III,
and Duncan Macdonald,
my best-ever field assistants

Contents

Acknowledgments

Gems and Minerals Magazine has kindly allowed me to use portions of several articles, including "Hobbyists, Amateurs, and Professionals" and slightly modified versions of "Techniques of Fossil Collecting," "A Fossil Lab on your Card Table," and "The Preparation of Fossils" as well as "Here There Be Sea Monsters," "Fishing in the Great Lakes of Wyoming," "The Hand Beast of Meteor Crater," and "The Great Midwestern Ocean."

Terra, formerly the *Los Angeles County Museum of Natural History Alliance Quarterly,* has allowed the reprinting of a slightly modified version of "The Maricopa Brea."

Masako Koda and Steffi Martens did the drafting and drawings.

Photographs of fossils were taken of specimens from the Department of Geologic Engineering and the Museum of Geology at the South Dakota School of Mines and Technology.

Introduction

Fossil hunting is an activity that almost any able-bodied person can do. The biggest restriction is living in an area where the rocks are not fossiliferous. However, weekend outings or vacation trips can generally take you to a place where you may prospect for fossils with a good chance of finding something worth collecting and taking home.

In this book I am going to try to cover the basics needed to properly collect fossils, record the data that will help you find something more than a curiosity, prepare the specimens you find for storage and display, and finally, organize a large field trip on which you might have a dozen or more participants.

Some years ago I published an article entitled "Hobbyists, Amateurs, and Professionals." I think that it expresses a philosophy that most of us who have spent our lives studying fossils as paleontologists would subscribe to. In the hope that it might be useful to you, it is reprinted below with the kind permission of *Gems and Minerals Magazine:*

> Years ago it was thought that a Royal Road to the understanding of evolution had been found. This idea was professorially stated as "ontogeny recapitulates phylogeny." This is *Sciench* for "the development of the individual retraces the development of the race." (This really shows that speaking in *Sciench* is sometimes simpler than using English, if you know the language.) In evolution this concept has been overthrown but, in the world of scientists, the title of this article describes both a phylogeny and, for some individuals, an ontogeny.
>
> It wasn't too long ago that there were no scientists. Societies weren't affluent enough to support them, and there were no good openings for an ambitious young man on his way up. At first, the person who dabbled in anything outside of the mundane things required to make a living was a hobbyist. He (and in all that follows, include "she") may have collected and speculated on rocks, minerals, fossils, sea shells, planets and stars, or any of the many other things that make up our universe.
>
> In ancient times the tenders of flocks used their nocturnal lonely vigils to study the stars, and it is still amazing how much true

astronomical knowledge they were able to acquire and coordinate. Eventually learned men of their time, usually doctors, lawyers, priests/ministers, or physicians and sometimes even wealthy and/or noble young men who were long on brains and bored with the usual pursuits of the wealthy and titled of the time, used their spare time to go a step beyond the hobbyist. These were the first amateurs and, at that time, all scientists were amateurs.

The amateur (A) today goes beyond the hobbyist (H). He makes a real study of the things he is interested in, reads profound and learned books, takes classes if available, becomes familiar with the professionals who work in museums or universities. He makes contributions to the scientific journals. Usually, the only difference between the serious amateur and the professional (P) in the sciences is that the amateur does something else for a living and probably doesn't have the "union card" to professionalism—the Ph.D.

In many areas the H and A don't create a problem for the P, except possibly for an occasional fractured ego. In astronomy, the H goes out in his backyard, sets up his telescope, and admires the craters of the moon or the closer view of the planets. The A spends hours in his home observatory with his homemade telescope scanning the heavens with a purpose and often makes new discoveries that the P might never make because his time is so taken up with scheduled problems.

It is only in the collecting sciences that there may be a real conflict between these various grades of interested people. This is particularly true in the collecting sciences that deal with non-reproducing things. The butterfly, shell, or egg collector can be reasonably sure that next year there will be another batch of these things to collect if the pressure doesn't become too great. In the non-reproducing field, it is usually a problem; often each item is unique or in very short supply. Also, it is of the greatest importance that the item's location in time and space and its relationship to other items be carefully recorded, or a potentially important scientific specimen becomes only a curiosity.

It is in the digging sciences that conflicts usually arise between the amateur and professional on the one hand and the hobbyist on the other. Always the P, and generally the A, believe that scientifically important material should be handled with respect, knowledge, and tender loving care. They have been given ample reason to suspect that, often, the H is interested in the specimen only as a curiosity and a trophy and, worse, that he believes he may collect it in any way he sees fit and has as much right to it as anyone.

There is no denying the fact that, under the law, all have equal rights to legally collected specimens, which usually includes the obtaining of governmental and/or private permission. The question is, does the taking of a scientifically valuable specimen for a private collection without regard to proper collecting methods have any ethical justification at all? As a P, I would say no; an A generally would lean rather strongly my way; many, if not most, H's would go along with us. Unfortunately, there are still too many who care nothing about the scientific value of a specimen if it will make a handsome addition to the private collection.

What do we do to get all of these conflicting interests together? Educational programs by the Federations and the individual clubs probably are the best approach. In such programs, they can always call on the local scientists to talk to and with the clubs in order to develop mutual understanding and rapport. The professional will teach them how to protect, collect and record the important geologic and geographic data on individual specimens. The clubs must impress on their members that museum-worthy specimens belong in museums and not in private collections. Private collections have a tendency to evaporate with changing interests or death. Museums hold things in trust for future generations, and many fossil specimens are unique records of life in the past.

An important concept is the respect for the professional's localities. I no longer publish locality data in my scientific papers. I have seen too many cases where non-professionals have gone to my localities, publication in hand, and collected from them. Their answer to a challenge on this activity is, "Well, you've published, so you must be finished with it." This is not true; we often return to localities after time has exposed more material for further collecting. Other museums also want reference material from these localities, and the ethics of the game are that you ask the original worker if you may collect from his sites.

I know one dealer and private museum owner who has been sending anyone who asks to one of my prized Miocene gravel pit sites, which lies about ten miles off the paved highway; the place is no longer worked for gravel, but it produces new fossils each year as wind and rain erode the surface. A paleontologist friend of mine was checking the site one day a few years ago and arrived just in time to meet a teenager walking out with a nearly complete rhinoceros skull!

At a recent mineral and rock show, an outstanding fossil exhibit consisted mainly of skulls from several of my late Oligocene localities in South Dakota. This material was collected from an

Indian Reservation without permits, allegedly for the geology department of a major university. Is there really any wonder that professionals many times develop strong anti-rockhound feelings? The condemnation of a large group on the basis of a few individuals is patently not fair, but we are all members of the same species and we all recognize that we have deliberately tried to control this tendency in ourselves.

Several times during the past few years, I have put my typewriter on my soap box and expressed my opinions on the ruination of important scientific specimens by uninformed or uncaring hobbyists. Generally the reactions from the non-P's has been very good, and the articles have been reprinted in various club and museum publications as far away as New Zealand. There have been a very few who cried havoc and seemed to feel that lynching was too good for me. As is quite understandable, unfavorable reactions are to be expected when the innocent feel they are being grouped with the guilty.

In this world of affluent societies, filled with people whose interests lie somewhere beyond the TV and the ball games, it is a natural thing for the energetic and intelligent ones to go out into the wonderful world of nature to observe, study and collect. Under these circumstances it becomes the duty of the organized groups to instruct, guide and educate. The professional will help the non-P if the latter is really interested in learning the scientific approach.

I've used the terms hobbyist, amateur and professional throughout this article. Why don't we make an acronym from H–A–P—Happiness Approaches Professionalism.

For those who care enough and who have developed the true amateur status, what can you do if you do find a fossil vertebrate, how can you determine if it is scientifically valuable, and preserve it for a museum? Follow these few hints and you will have your problems well in hand:

1. Take a picture of the find if it can not be picked up without damage.

2. If it is very fragile or broken, soak it in the ground with thin white shellac. Pick up the pieces that have been washed down the hill and pile them with the parts still in place.

3. Cover the specimen with newspaper or plastic and pile dirt on it. If you put a note or a card in a bottle with the specimen, it will sometimes prevent vandalism if found by a second party before you return.

4. Mark the spot so you can find it again.
5. Record all the geographic and geologic data you can get.
6. Send all of this information to the nearest vertebrate paleontologist. Only an expert can tell if it's scientifically valuable.
7. If you don't get an answer within a week or so, harass him a little. Remember, in the summertime he probably will be on a field trip.

Most major universities and museums have paleontologists, and most of them are courteous and interested enough to answer your letters. The paleontologist will tell you if your find is of scientific value and museum-worthy. If so, you have made a contribution to science; if not, you have a specimen for your collection with the knowledge that you are not depriving science and future generations of scientific information. This could be a road to immortality. If the specimen represents a heretofore unknown animal, the scientist may name it for you. If he does, as long as science exists on this planet, your name will be in its literature.

Chapter One

THE ORIGIN OF FOSSILS: NATURAL UNDERTAKING

Most people think of a paleontologist as someone who studies fossils. That is true, but only in a limited sense. The paleontologist does study fossils, not as objects, but rather as symbols representing either the remains or evidences of the activities of ancient organisms. He is primarily interested in reconstructing the plants and animals of the past, determining their evolutionary development, their comings and goings across the globe, reconstructing ancient plant and animal communities, determining the habitats, and establishing the relative ages of the rocks in which these organisms lived. Fossils are the clues used to solve these mysteries.

WHAT IS A FOSSIL?

Fossils are commonly thought of in terms of petrified bones, shells, and logs. Unfortunately, the processes that preserve or petrify are referred to as fossilization. Actually, a fossil is a state of being, and the process that forms the fossil is that of preservation. Fossils may take many forms and result from a number of different processes, but they all have one thing in common. They are the remains or traces of organisms that existed in past geological time.

Fossils result from a sequence of unusual events which preserve only a minute fraction of all the organisms which lived in the geologic past. The conditions under which preservation can take place are so exacting that only a few individuals in any plant or animal community can qualify as potential fossils.

HOW FOSSILS ARE PRESERVED

I. Conditions favoring preservation.
   A. Presence of hard parts—bone, shell, woody tissue.

B. Rapid burial.
C. Burial in fine sediments.
D. Low bacterial association.
E. Uniform conditions of temperature and humidity.
F. Circulation of ground water carrying dissolved minerals and having an acidity not hostile to the chemical composition of the remains. (Fossils are *rarely* preserved in igneous or metamorphic rocks.)

II. Modes of preservation.
   A. Without change.
      1. Freezing in permafrost.
      2. Desiccation in desert caves.
      3. Impregnation with asphalt.
      4. Buried, but unchanged mineralogically.
      5. Insects in petrified tree sap or amber.
   B. With change.
      1. *Permineralization*: original hard parts remain, with mineral matter deposited in pores, cavities, or voids.
      2. *Replacement*: virtually complete or nearly complete replacement of organic material by waterborne minerals. Silica, calcium carbonate, and calcium phosphate most common; pyrite and carbon approximately 10 percent; but any waterborne mineral may occur.
      3. *Carbonization*: reduction of soft parts to carbonaceous films; also coalification.
   C. Molds and casts.
      1. "Two-dimensional": imprints of leaves, feathers, and delicate organisms.
      2. Three-dimensional: natural molds and casts in sediments and lavas. These include external molds, internal molds (or cavity fillings), natural casts by sediments or precipitates, and insects in amber which are preserved as a combination of molds and unaltered tissue.
   D. Activity indicators.
      1. Tracks, trackways, and trails.
      2. Burrows in soft sediment.
      3. Borings in hard rock.
      4. Tooth marks and other evidence of feeding.
      5. Coprolites (petrified or unaltered excrement).

This means that there are great gaps in the fossil record—perhaps only a few members of a group preserved, and many groups, certainly, from which none were preserved. Some habitats, such as those on high mountains where the land is constantly being worn away by erosion, go completely unrecorded because there is no opportunity for burial or other means of preservation.

CONDITIONS FAVORING PRESERVATION

There are six primary requirements for preservation. As with most rules there are exceptions, but these will cover the great mass of preserved fossils.

1. *The presence of hard parts.* The good fossil record begins about 600 million years ago when organisms began to develop hard parts for support or protection—shells, and later bones and woody tissues. Soft tissues such as the flesh of animals are rapidly destroyed by predators, scavengers, bacteria, and the physical environment, so they seldom last long enough for the various processes of preservation to do their work.

2. *Rapid burial.* When an organism dies it is immediately set upon by scavengers, bacteria, wind and rain, stream currents, or ocean waves. Burial under a blanket of sediments will greatly restrict this destruction if it happens quickly enough and give the processes of preservation a chance to act on the remains. Plants and animals living in areas of erosion are rarely preserved. Only in special circumstances, or if the remains are washed into a local area of deposition such as a lake or valley bottom, are highland organisms preserved. For example, in the White River Badlands of South Dakota certain stream channel deposits contain the battered remains of *Protoceras,* a small horned mammal that was obviously not a river dweller and was apparently adapted for life in the uplands. As these channels originated in the Black Hills to the west, it seems that occasional *Protoceras* corpses were carried downstream to be buried in the channels as the streams slowed down on the flats below.

3. *Burial in fine-grained sediments.* Clay-size particles, silts, and sands are the best sediments for the burial of potential fossils. Such fine-grained particles gently encase fragile remains without abrading them or pounding them into fragments. It is in the very finest sediments that we occasionally find indications of organisms which had only soft parts. Fine impressions of such soft-bodied organisms as jellyfish may be found between layers of clays and fine silts. When such fossils come to light they are a welcome and unexpected glimpse of a little-known aspect of past life.

4. *Low association of bacteria and other destructive biologic agents.* Bacteria are among the saprophitic agents of decay that quickly destroy

the soft parts of dead organisms and eventually many of the hard parts as well. Burial sometimes excludes bacteria, yeasts, and molds, as will dessication in a desert environment, freezing in arctic sediments, or being coated with an antiseptic material such as petroleum.

5. *Underground water carrying dissolved minerals.* Not all fossils are "petrified," but those that are get their stony material from mineral-carrying ground water that percolates beneath the surface of the earth. As ground water moves through joints, cracks, and pores in the rock and sediments, it takes ions of soluble minerals into solution to be precipitated later at some other place. The scale that develops inside your teakettle is such a precipitate. As the water moves through buried bone or other material, it may deposit any of the ions it carries during the process of petrification. Sometimes even radioactive minerals are left behind; this led to the wholesale destruction of many fine fossils by uranium hunters in the 1940s and 1950s when the uranium hunt was in full cry.

6. *Little variation in temperature and humidity.* Uniform conditions are generally required for good preservation. Repeated freezing and thawing or wetting and drying will break up a potential fossil and either destroy it completely or greatly reduce its scientific value.

Given all these conditions a specimen may be preserved. Yet even after meeting them all, there is a great probability that the resulting fossil will never come to light. It may be destroyed by later erosion. If not destroyed, it may be so completely buried that it will fail to come to the surface for us to collect it. If it is exposed at the right time, will its collector recognize it for what it is? Will it fall into the hands of a paleontologist who can classify, identify, and study it? The odds against each of these conditions are so high that it is a wonder we know as much about the life of the past as we do.

TYPES OF PRESERVATION

Fossils may be preserved by very different processes. Different sorts of organisms, preserved under varying conditions, may come down to us in a variety of states. Some fossils are virtually unaltered by the preservation process; others may have no part of the original organism remaining.

Preservation without any physical change or with very little change is of great interest; we can learn about the internal organization and body chemistry of such organisms or at least study the structures without the interference of foreign materials. Occasionally whole carcasses of animals have been preserved in the frozen soils of the Siberian and American Arctic. The most publicized of these was the discovery of the famous mammoth from near the Berescovca River, well within the Arctic Circle, in northern Siberia. It was only one of a number of such finds, but this one was eventually collected in 1900 and mounted as both a skin and a skeleton in the museum at St. Petersburg (now Leningrad). Contrary to a few sensational articles which were published about this find, there were no "mammoth steaks" served at banquets (one account by a member of the collecting team mentioned that the beast could be smelled several kilometers away), there were no fresh buttercups preserved in the mouth, and an autopsy revealed that the animal had died of suffocation, not by instant freezing in a cosmic disaster. There, in a place where the temperature seldom rises above freezing, the Berescovca mammoth had apparently fallen from a cut river bank into the muddy stream, and before long he died of his injuries and suffocation while struggling in the mud.

(Some years ago the author of a sensational article about the fate of the Berescovca mammoth also published an article about a frozen primitive man from Wisconsin in a popular men's magazine. It was a fine story with no end of scientific documentation and evidence for this and that. The best part of the story was that the ancient man was made by a friend who was a talented sculptor in Southern California).

Dry desert caves have produced large numbers of bones and sometimes even flesh, skin, hair, and dung of late Ice Age or Pleistocene animals. These remains are unchanged chemically although the carcasses have been thoroughly desiccated and mummified in the dry desert air. These fossils may represent either animals which lived in the cave or the scattered refuse of hunters and scavengers, including man, who dragged their meals into an underground lair. A number of such caves in the southwestern United States, California, southern Nevada, and northern Mexico have been the source of vital information on the late Pleistocene flora and fauna of these areas.

Peat bogs and asphalt seeps are also sources of unchanged bones and plant material; under very favorable conditions even flesh and hair may be preserved. Animals lured into these natural traps by the film of water on the surfaces became stuck and then acted as bait to attract other animals. Carnivores, attracted by the cries of the already trapped animals, would in turn become trapped and preserved as fossils. Southern California has four large asphalt traps which have yielded thousands of specimens. The most famous of these is Rancho la Brea near the heart of Los Angeles.

Occasionally shells will be preserved without any change when buried in sediments. In this case the ground water has neither added nor subtracted from the original material.

Preservation with chemical change usually involves some degree of mineralization through the action of ground water. In some cases wood may be "coalified" or carbonized, but usually there is mineral addition or replacement. Silica, calcium carbonate (calcite), and calcium phosphate (apatite or collophane) are the most common replacement minerals and are found in some 90 percent of mineralized (petrified) fossils. Iron sulfide (pyrite) and carbon are found in most of the remaining 10 percent. Many other minerals may be involved in preservation, but overall they are very rare.

If a fossil is only partially replaced, the open spaces filled with minerals, and much of the original material intact, as is the usual case with fossil bones, we say the object is permineralized. In cases of complete replacement—for example, agatized wood—we say the object is petrified, and the process is called petrification. There is no defined boundary between the different end products, and the processes, of course, grade into each other.

Many fossils may be only the impressions of organisms. If a bone or shell is buried and then dissolved away by ground water, it may leave a cavity in the rock which forms a natural mold. Sometimes we find just the mold and sometimes we find one which has been refilled with other minerals by the ground water, forming a natural cast. This is a three-dimensional replica of the outward form of the organism. The delicate insect fossils preserved in amber, a fossil resin, are actually combinations of original tissue and natural molds. Hard external parts such as wings, carapaces, and antennae remain stuck to the insides of the molds, although most of the bodies are withered away. When molds of larger organisms have been preserved without subsequent filling, a paleontologist may fill the cavity with plaster, plastic, or liquid latex and make a cast from it. Often we find molds of snails or clams in which the shell was both surrounded and filled with sediments before it dissolved. A thin space separates the internal and external molds where the shell once was. In the casting industry the internal molds are called core molds, but in paleontology they are often referred to as steinkerns (literally "stone kernels").

Leaves, feathers, insects, and soft-bodied animals are often preserved as flat, two-dimensional impressions on fine-grained sediments. The famous transitional reptile/bird *Archaeopteryx,* found in a limestone quarry in Bavaria, would probably have been classified initially as a small dinosaur if two-dimensional imprints of its feathers hadn't been found around the wings and tail.

Indications of animals' activities, sometimes called "lebensspuren" by paleontologists, are often preserved. We usually refer to these as "trace fossils." Tracks, trails, boring, burrows, feeding struc-

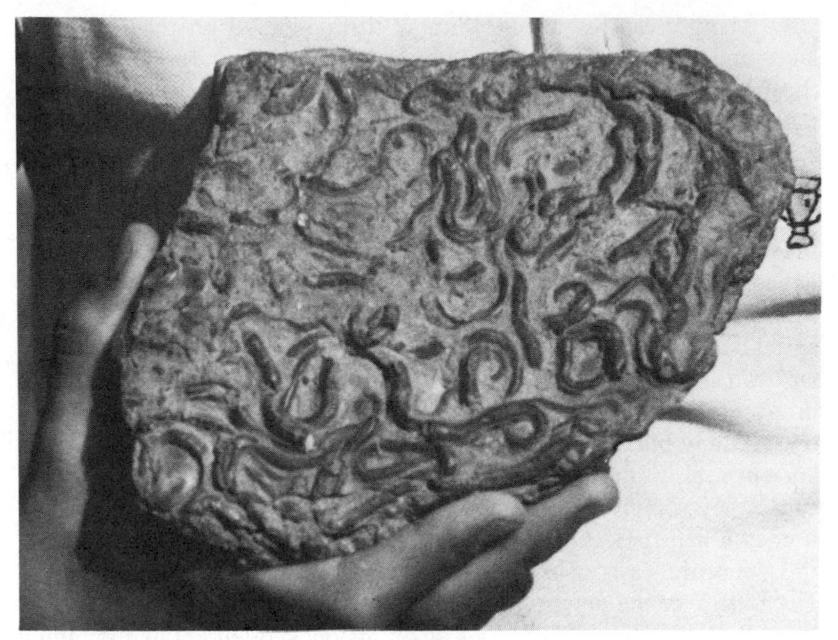

Figure 1. Worm borings. These are known as trace fossils or lebensspuren.

tures, tooth marks, resting imprints, and even preserved feces are all activity indicators. Often the problem is in deciding what made the structure or tracks. Footprints of camels, horses, dogs, and cats are easily matched to the proper sort of animal because their relatives are still living and making new tracks to compare to the ancient ones. For extinct groups with no close living relatives, difficulties arise. Many of their tracks tell us a great deal about the animals that made them, yet we are at a loss to decide which skeletal remains belong with the tracks. For example, there are footprints in the Triassic Moenkopi Formation in Arizona deposited 200 million years ago in red sediments laid down on a broad subtropical or tropical coastal plain during the early part of the Age of Reptiles. These show that the local amphibians, big salamanderlike creatures up to 2 meters in length, wandered aimlessly about the mudflats and muddy river banks. The reptiles, which left a peculiar hand-shaped print, seemed to be striding purposefully either up- or downstream as they went about their reptilian business, probably eating amphibians. We have a fair idea of which skeletons would match the amphibian footprints; but while the reptile footprints have been extensively studied and sound deductions made about the animal that produced them, to date no skeleton belonging to such an animal has ever been identified.

Many creatures which winnow through mud on the bottom of lakes and seas leave meandering trails showing their routes across or through the bottom. A freshwater mud flat early in the Age of Mammals, in the Eocene epoch, records the passage of a flamingo which left both footprints and the trace of its bill as it winnowed the mud of a lakeshore in search of tasty tidbits.

Rock-boring clams often leave both their holes and their shells preserved in the rocks of ancient shallow seas. In early Miocene rocks deposited about 25 million years ago during the middle of the Age of Mammals in western Nebraska and southeastern Wyoming, we find the spiral burrows of beavers who lived like prairie dogs preserved in regular town sites. At Agate Fossil Beds National Monument in Nebraska an interpretive exhibit and trail through such a townsite is being developed. Many burrows are exposed on the walls of canyons cut into the ancient prairie surface. We know the beavers made these burrows because they sometimes left their skeletons in them when they died at home. When these burrows were first discovered it was thought that they were the casts of the taproot of a giant plant. They were named *Daemonelix* or the devil's corkscrew.

Preserved feces (coprolites or dungstones) often tell us something of the feeding habits of fossil animals. Shortly after World War II the late Dr. R. A. Stirton of the University of California discovered a large fox-sized carnivorous marsupial skeleton in Miocene rocks in Columbia. When the skeleton was prepared in the laboratory in Berkeley it was discovered that in its death throes the marsupial had evacuated a mass of excrement. In this mass of coprolite (fossil dung) was found part of the skull of a small rodent. Three bits of information for the price of one: a new sort of carnivorous marsupial, another record of a contemporaneous rodent, and the information that the marsupial's diet was at least partially based on rodents.

Great accumulation of ground sloth dung in a dry desert cave northeast of Las Vegas, Nevada, showed that these animals lived largely on cholla and prickly pear cactus. If you can relate the coprolites to the producer, a great amount of extra information on life styles may be available.

EXCEPTIONS THAT PROBE THE RULES

Most fossils are preserved in sedimentary rocks. Metamorphic rocks (those changed from their original state by great heat and pressure) may have contained fossils at one time, but the process of change and recrystallization usually destroys any preserved fossils. Fossils are rarely found in igneous rocks, those produced from the molten state, but some exceptions exist. Deposits of tephra, material that is blown from a volcano and cools in the air as it settles to the ground, may contain fossils. Very fine dust or sand-grain-sized particles may be carried for miles by the wind and accumulate on the ground in thick blankets like a fine-grained sediment. A dramatic collection of dozens of complete skeletons of the Miocene rhinoceros *Teleoceras* and other mammals and birds was recently excavated from such deposits in northeastern Nebraska where they had been smothered by windblown volcanic ash.

In rare cases fossils have been found in lava flows. A very dramatic example is the Blue Lake rhinoceros from the Grand Coulee on the Columbia Plateau in eastern Washington State. In the summer of 1935 a hiking party found a small cave about the size of an office desk, with a man-sized entrance. One of the men crawled into the cave and emerged with some bone scraps and a jaw fragment with broken teeth. Eventually the bones were given to George Beck, a fine amateur collector who taught geology at Central Washington State College in Ellensburg. Beck sent some of the specimens to the California Institute of Technology where they were identified as belonging to a rhinoceros. A short paper by Beck suggested that the cave was actually the natural mold of a rhinoceros. Some claimed it would be impossible for a rhino to be molded in a lava flow. It was even suggested that the bones had been carried into the cave by a coyote. Since there are relatively few rhinos in North America these days, there had to be a better explanation.

After World War II the original bones and some additional material collected by Beck and others went to the University of California at Berkeley. A group from the Museum of Paleontology decided to settle the matter once and for all. They went to the site, cleaned out the cave, and found some more foot bones in the process. They then greased the inside of the cave and "papered" it with plaster and burlap. When the lining had hardened they cut it loose and reassembled it outside the cave. When it was done they had a replica of the bloated carcass of the early Miocene two-horned rhinoceros *Diceratherium*. It lay on its side, legs extended, the body bloated by the gases of decay. The entrance of the cave was at the buttocks, the middle of its back was against the mold of a tree trunk, and the lava was in pillows which form when the lava flows under water.

Reconstructing the story was now quite simple. The rhino, a common early Miocene species, had died and the carcass become bloated. It was washed downstream sometime during this period and came to rest against a tree. Possibly he (the horn bases were preserved, and females of this group do not have horns) was killed by gases from a volcanic eruption; but whatever the cause of death, lava soon flowed down the dammed-up stream and covered the body. Rapidly chilling in the water, the lava pillowed up over the carcass, forming a natural mold. Over the years lavas continued to build the plateau. Millions of years later the Columbia River, fed by melt water from retreating Ice Age glaciers, carved the Grand Coulee and in the process sliced off the rump of the rhino mold. Perhaps the lesson to be learned from this story is that not only are fossils sometimes preserved under amazing circumstances, but that one should never say "never" until all the facts are in.

Under very special conditions other than freezing or dessication soft parts of organisms may be preserved. Some of the most spectacular examples are the middle Eocene fossils in the oil shales at Messel in West Germany. Here entire flattened bodies of small horses and other mammals are preserved with hair, skin, flesh, and even stomach contents. It is incredible to think that such structures could have withstood the vicissitudes of 58 million years of earth history.

Chapter Two

WHERE TO LOOK FOR FOSSILS

WASTELANDS AND BADLANDS

If you are traveling through the arid parts of the western United States you are in the heart of excellent fossil hunting country. Western Kansas is famous for the late Cretaceous Niobrara Chalk which contains fish up to 12 feet long. One such monster was a double find as he had gulped down a 6-foot friend just before he died—possibly as the result of his greed. There are also marine turtles, marine lizards (mososaurs), long-necked plesiosaurs, and even a wingless bird up to 5 feet long who lived like a seal. There are also numerous invertebrates, including nautoloids and ammonites.

In the Dakotas, Nebraska, and eastern Colorado the chalk is not as fossiliferous, but the same fossils may be found in the dark shales which overlay and underlay the Niobrara Chalk. All of these beds were laid down in a shallow seaway that extended from the Arctic Ocean to the Gulf of Mexico.

Because of the arid climate, soils have not developed in many areas, so the sediments in these wastelands are exposed and readily prospected.

In western South Dakota, western Nebraska, and northeastern Colorado extensive areas of badlands have developed in flat-lying Oligocene and Miocene sediments. Here some of the world's finest collections of fossil mammals have been made. Fossils from these beds are on display at the Nebraska State Museum in Lincoln, the Denver Natural History Museum, Agate Fossil Beds National Monument in western Nebraska, the Badlands National Park in western South Dakota, and the Museum of Geology at the South Dakota School of Mines and Technology in Rapid City.

Wyoming and Utah have a number of intermontaine basins that were formed when the Rocky Mountains were uplifted at the end of the Mesozoic Era. There are both lake- and land-laid sediments in these basins. The badlands of the Bighorn Basin in northcentral Wyoming is a particularly interesting place to prospect. Check with the Bureau of

Figure 2. Collecting a brontothere skull and a rhinoceros skull in the White River Badlands of South Dakota. The sunscreen is an exercise in futility as the wind had it down in a few minutes.

Figure 3. A beautifully preserved fish from the Eocene Green River lake beds in Wyoming.

Land Management Office in Worland to determine if permits are required on the public lands.

Wyoming's Green River Basin is famous for its beautifully preserved Eocene fish. There are plants, insects, stingrays, paddle fish, gar, bowfins, herrings, catfish, perch, turtles, crocodiles, birds, bats, and other mammals preserved in these fine-grained sediments. Be sure to visit Fossil Butte National Monument near Kemmerer when you are in the area.

STREAM AND RIVER CLIFFS

The cliffs formed along the courses of rivers and streams are excellent places to prospect wherever sedimentary rocks are found. The cliffs along the rivers of the upper Midwest are often cut in fossiliferous limestones which are rich in invertebrates such as brachiopods, crinoids, and many other beautiful fossils.

The cliffs along the Missouri River are often rich in marine rep-

Figure 4. The writer digging for nodules along the Snake River in Idaho.

tiles which are constantly exposed as the sides of the cliffs slump away as the result of undercutting by the river.

Where the Snake River in southern Idaho cuts through sedimentary rocks you can find nodules containing crayfish constantly weathering out of the soft sediments.

WAVE-CUT CLIFFS

Wave-cut cliffs along lake borders and sea coasts are always excellent prospecting areas when they are cut in sedimentary rocks. The cliffs develop as the waves cut a notch in the shore at high tide level. The undercut portion of the shore breaks away and falls to the beach, creating a cliff. Thus the cliffs are constantly retreating and new material is being exposed. Extreme care should be exercised in working these cliffs as they may fall at any time.

South of San Francisco at Montara Beach the cliffs have retreated nearly a quarter of a mile in the past 50 years. This is probably the result of the suburban growth on the San Francisco peninsula. Pollution from these developments has been carried southward by the longshore currents. This pollution has killed the kelp which formerly formed thick mats a few hundred yards offshore. With the kelp beds no longer dampening the force of the waves, the force of the attack on the shore has increased many times.

On the East Coast the Calvert Cliffs of Maryland are an excellent place to prospect as they are particularly rich in shark teeth.

One plus for the prospecting of sea cliffs is that the tide goes out twice a day and they are therefore easy to get to.

BREAS OR ASPHALT DEPOSITS

In many parts of the world fossils have been preserved in asphalt seeps. Perhaps the most famous of these is Rancho la Brea in Los Angeles, California. There are a number of other breas in Southern California as well as in South America. The following article reprinted from the Los Angeles County Museum of Natural History's *Quarterly* tells about the discovery of a brea deposit by an amateur collector and its subsequent excavation by museum crews.

THE MARICOPA BREA

> The name of this fossil deposit sounds like a fancy nightclub, and when we add the name of the finder it becomes Wally Block's Maricopa Brea—more nightclubish than ever. Actually, the Mar-

icopa Brea is a late Pleistocene deposit of fossil animals, closely related in time and composition to Rancho la Brea.

Several years ago, W. E. Block, an engineer with the Standard Oil Company of California, found this deposit while stationed at Taft, California. He collected a few specimens and brought them to the Museum for identification. He was encouraged to continue collecting but was then transferred to San Francisco. Upon reassignment to Southern California, he resumed his collecting activities and donated a large number of specimens to the Museum. A couple of such donations led to an examination of the site by the writer and other members of the staff. Wally had found a fossil treasure trove.

It seemed for a week or so that we were commuting back and forth between Los Angeles and Maricopa. The site was a magnet that pulled us as often as we could get away. Sabre-tooth cats, gigantic jaguars, a bear larger than the Kodiak, dire wolves, coyotes, badgers, eagles, vultures, rabbits, horses, camels, bison, meadowmice, snakes and beetles were the initial take.

Here was the fossil find of the decade. Preliminary studies suggested that this was an ancient waterhole near the foot of the

Figure 5. The Maricopa Brea in Southern California.

mountains. The San Emigdio Range rising from the plain about a mile to the south probably looks much the same today as it did when the fossils were being accumulated. To the north stretches the gently sloping alluvial apron of the southern end of the San Joaquin Valley. With a slightly moister climate producing more trees and taller grass, the local scene differed little in the late Pleistocene from what can be seen today. The immediate site of the locale is the south edge of a very low, broad hill rising not more than twenty feet above the plain. The hill is capped by about two feet of asphalt. Whether this is a natural occurrence or the result of dumping from a shallow oil well that was dug here, is not known at the present time. At the southern foot of this mound is a series of shallow sag ponds, small depressions developed along a fault zone, dry now, but probably briefly filled with a foot of water after rainy spells, and certainly the site of springs in the recent geologic past. The bone bed lies between the sag ponds and the base of the slope. A foot to two feet in thickness, it is covered with a minimum of two feet of oil impregnated sand, silt and clay overburden. The bones are found in a mixture of clay and asphalt. Unlike Rancho la Brea, they are not heavily impregnated with petroleum and asphalt but are "dry" and brittle like the bones found in many cave deposits. There is little association of skeletal remains. Most of the skeletons have been badly scattered and mixed into a pile of jackstraws.

As soon as a site is found, the investigator immediately begins working on hypotheses to explain the origin and mode of accumulation of the deposit. "The Scientific Method" comes into play. This is a somewhat stuffy way of saying that you are trying to logically explain a natural phenomena. There are basic steps to follow in this reasoning process, but I doubt very much if anyone ever really formalizes their thinking into these phases in the way one would when writing the formal propositions for a debate. Basically, the scientific method attempts to answer the following questions:

- What are the known facts of the problem?
- How do these facts appear to be related?
- How may this relationship be explained?
- Is this explanation supported by all obtainable new information?

When this process is "completed" you have developed a hypothesis. Then the sniping begins. Others will have arrived at different answers resulting from the application of their training and background. If your shirt isn't too badly stuffed, you will welcome these ideas and actually encourage someone to act as the "Devil's Advocate" in order to pull down your hypothesis. This is part of the great fun ot science. Sometimes scientists, who are also people, get emotionally involved with their hypotheses: they consider them as their children who must be defended from the onslaughts of the unenlightened who look upon them as little monsters. It is unfortunate that such differences of opinion have caused lifelong feuds and have developed great schisms within disciplines. Some of these splits have not really healed after the passage of several generations. One of the great feuds in this country in Vertebrate Paleontology developed between E. D. Cope and Othniel Marsh during the last third of the 19th century. Both were great men who were pushing back the frontiers of knowledge and stepping on each other's toes while they were doing it. Even today, you will find students of the protagonists "who would rather fight than switch," or even concede a modicum of greatness to the opponent.

A mature worker will accept the ideas and concepts of others to be weighed against his hypothesis; some such thoughts may be incorporated into his scheme and others rejected. Friendships are not shattered, although you may reserve a small corner in which you feel that your friend and co-worker of many years is, in this case, obviously blinded and showing either signs of idiocy or incipient senility.

The principle of Occam's Razor must always be kept in mind: the simplest explanation generally turns out to be the most logical. Complexly structured hypotheses collapse with an embarrassing crash when a well-administered logical kick is applied to an improperly balanced fundament.

So, the question is, how did the Maricopa Brea deposit develop? Preliminary working hypotheses might include, (1) natural deaths at the waterhole, (2) entrapment in the mud at the waterhole, (3) entrapment in the asphalt at the site, or (4) a combination of these. The asphalt associated with the material does not appear to be a trapping factor but simply a preservative which kept the ground water and bacteria away from the bones. Further studies and population analysis may change this concept: the abundance of carnivores does present a problem at this writing and will have to be fitted into the final solution.

Finding a site and making short collecting trips with staff and volunteers is one thing, but finding money, people and time for proper collecting is something else again. County travel funds are practically nonexistent. The members of the laboratory crew in the Vertebrate Paleontology Section are virtually chained to their benches with our expanding exhibit program. When each mounted dinosaur skeleton represents from two to five years of preparatory time in the laboratory, four preparators don't stretch very far.

Fortunately in this case, three necessary ingredients for a successful exploitation of a site came together. A concentration of fossil material has been found and made available to us. Financing is being supplied by an anonymous donor who has generously supported two other fossil digs and is always interested in new and exciting discoveries. An available collector was hired. . . . More than a dozen plastered blocks of fossils were removed and a great number of small bags of isolated bones and skulls came with them.

Members of the regular Museum staff continued to make at least weekly visits to the site to bring supplies to the field, help with the plastering and moving of large blocks, and to haul the loot back to the Museum.

When Jimmy's month ran out we hired Wally Block's nephew, college student James Farrell of Taft and a friend to continue the work through the summer. The number of specimens constantly increased and a few new animals were found. It is much too early to give a complete list of the fauna as several years preparation will be needed before all of the material is removed from its blocks and prepared for study.

The fauna is similar to that found at Rancho la Brea and the brea deposit at McKittrick about thirty miles west of Maricopa. A preliminary review suggests that some of the carnivores, particularly the dire wolf and sabre-tooth cat, are somewhat larger than those from Rancho la Brea. The study of this fauna and a comparison with the Rancho la Brea and McKittrick faunas will give science an important insight into local population variations during the late Pleistocene in Southern California.

One unpleasant aspect of the work was the presence of Valley Fever in the area. This debilitating lung condition is caused by spores present in the dust and dirt. Endemic in the San Joaquin and San Fernando Valleys, the oily dust of the Maricopa Brea put three of the commuting diggers to bed, with one spending a week in the hospital. Suddenly we were all guinea pigs for a medical group studying the disease. Blood was taken, tests were given and to protect the rest of the staff, all of our collections went through the Museum's fumigator. Well, that's Vertebrate Paleontology—you never know what your investigations are going to lead to.

WAVE-BUILT TERRACES

As waves cut cliffs along seashores and lakeshores the rock which falls to the beach is slowly carried to sea by the waves and deposited offshore as a wave-built terrace. When terraces are exposed by an uplift of the land or a dropping of the water level they become good areas for fossil prospecting. Besides invertebrates they may contain shark teeth, the bones of whales, sea lions, and seals, and even the remains of land animals.

Marine wave-built terraces are exposed on both coasts, and fossils are often exposed when builders are constructing homes with a view along the coast. I once went to such a development to look at a find made by the construction people which turned out to be a fragment of a gigantic whale skull. Of course, geologically speaking, those who buy homes on such terrace deposits by the sea are in need of geological education or a mental institution.

During the Pleistocene or Ice Age there was much more rainfall in the now-arid West. The basins of northern Nevada were filled with water, forming a many-armed lake which we now call Lake Lahonton. In Southern California there was a string of interconnected lakes filling basins extending from Owens Valley at the base of the Sierra Nevada to Death Valley on the Nevada border. All of these produced both terrace deposits and lake deposits which are in some places rich with fossils.

The most spectacular terrace deposits in this area are those which surround ancient Lake Bonneville. This lake occupied the Great Salt Lake Basin near the end of the Pleistocene. At its highest stand it was 1000 feet above the present surface of the Great Salt Lake. There are 50 or more terraces, but only 3 are important. These are beautifully developed on the Wasatch Front north and south of and under Salt Lake City. There are many gravel pits on these terraces which produce such creatures as horses, camels, and muskoxen.

There are hundreds of square miles of these terraces which deserve attention. James Madsen, the Utah State paleontologist, has a rapport with many of the pit operators, so they report finds to him so he can continue to add to the record of ancient life in Utah.

The Great Lakes were formed when the land was pushed down under the weight of thousands of feet of ice during the Pleistocene. The land is slowly rising as it regains its equilibrium. Wave-built terraces along the margins of these lakes should be prospected.

ROAD AND RAILROAD CUTS

In the more humid parts of the country, road and railroad cuts may be the best exposures available. Fresh cuts are quickly covered by vegetation, so you have to be quick.

Be careful when you are prospecting these cuts, and for goodness' sake don't include your children in the party.

Near Verdi, Nevada on the eastern base of the Sierra Nevada is a wonderful fossil plant locality in a Southern Pacific Railroad cut. Care has to be taken, as the modern diesels sneak up on you. Before World War II when the trains were hauled over the hill by great articulated

Figure 6. Prospecting for Miocene leaf imprints in a Southern Pacific Railroad cut near Verdi, Nevada.

engines with 12 6-foot drivers you could hear them coming as they roared down the hill going flat out. As they thundered through the cut only the hardy survived.

It is well to check with the police before prospecting cuts on major highways. I've had a couple of encounters with the California Highway Patrol when I've had classes working a cut on the interstate. Maybe we could do a "CHIPs" program on this!

CAVES

Caves are often a happy fossil hunting ground. Most caves develop in limestone, and as limestone formations are fairly common on this continent there should be plenty of opportunities for cave exploration. If you should discover a fossil cave deposit of any magnitude, the excavation should be done under the supervision of a professional.

Many western caves have deposits of ground sloth dung in them. In one such cave near Las Vegas, the partially charred bones of a small camel, ancestor of the llama of South America, was found with the dung. I've always wondered what camel tastes like when roasted over a fire of sloth dung!

Some years ago I went to Auburn, California to take pictures of a famous fossiliferous cave. When I reached the site I discovered that the limestone had been quarried for cement and the cave was gone. This is the same thing that has happened to the South African cave deposits where early man was first found.

Another fossiliferous cave near Mount Shasta in Northern California has a steel grid over the entrance because *Boobus americanus* had seen fit to fill it with beer cans and write its name on the walls with paint spray cans.

FISSURE FILLS

Sometimes when there is orogeny (mountain building), large cracks or fissures develop in the rocks. These are great depositories for later sediments. Digging out these fissure fillings and getting fossils from them is sometimes a bit of work. As I write this I've just finished with the "picking" of such a fissure fill. The crack was in the Pennsylvanian Pahasapa Limestone in South Dakota's Black Hills. We took out 500 to 600 pounds of matrix and washed it in screens in Rapid Creek in Rapid City. Getting rid of much of the clay and silt reduced the bulk to about 200 pounds. This was packed in plastic garbage bags in boxes and taken to California. Here I washed and screened it some more and got down to four 3-pound coffee cans of concentrate. After this it is a teaspoon at a

Figure 7. An Oligocene fissure fill in the Mississippian Pahasapa Limestone in the Black Hills of South Dakota. Six hundred pounds of matrix yielded a heaping tablespoon's worth of teeth and tiny bones.

time under-the-microscope picking job. One of my students picked, and so did I. It was about a two-person year to do it. What did we get? Twenty-five to 30 small mammal teeth, a few snake vertebrae, some lizard plates, and a few snails. What have we learned? A great deal about the uplifting of the Black Hills. Science is only spectacular on TV. Most of us are those who only sit and sweat.

LAKE BASINS

I've sort of indirectly touched on lake basins when I was talking about the intermontaine basins in the Rocky Mountains. Lake basins often produce beautiful fossils because of the fine-grained sediments that

Figure 8. Tertiary freshwater fish preserved in diatomite.

settle to the bottom some distance from the high-energy areas at the mouths of streams.

The Green River Basin in Wyoming contains beautiful fossils, as I mentioned above. But there are other lake beds that also contain some pretty nice fossils. I like to take my classes to a place called Two Tips south of Nevada where we get leaves, flying seeds, insects, and fish.

SAND AND GRAVEL PITS

Sand and gravel pits are found just about everywhere. They are always potentially good fossil sites. Cultivating a relationship with the pit operator could be rewarding in terms of fossil collecting. James Madsen, the Utah State paleontologist, has developed a fine rapport with the pit operators along the old lake terraces near Salt Lake City.

There are several ways to work a pit. If the operator is willing, you can follow the bulldozer, can, or dragline and see what is being uncovered. Some pits have conveyor belts running from the source to the screens and stockpiles. Sitting or standing by the belt gives you an opportunity to grab specimens as they pass by. I knew one pit operator whose wife set up by the belt with a chair and beach umbrella every day and made a fine collection of middle Pleistocene land mammals.

If the operator doesn't want you about when the pit is in operation you perhaps could get permission to prospect on weekends and holidays. Near the coast about 50 miles south of San Francisco, there are a series of sand pits several hundred feet thick and covering well over a square mile. This sand is unconsolidated and contains lots of shark teeth and bones of marine mammals. We were often allowed to take classes into the pits on weekends and screen for fossils.

Some years ago a pit was temporarily opened near Mission, South

Dakota for a highway construction job. The contractor brought to the museum some horse teeth which had been found in the pit. Sand and gravel were being bulldozed into a hopper where it was carried by belt up to some shaking screens. The desired material fell into waiting trucks below, and the oversize rocks and fossils slid down a discard chute. For the next several weeks I stood at the bottom of the chute with a garden rake, rescuing fossils and dodging flying rocks. By the time I got back to the motel each evening my lower legs above boot top and my hands looked as if I'd been in a fight with a thrashing machine. The final collection contained teeth, jaws, and various bones of several kinds of camels, horses, and dogs, as well as parts of a tapir, mastodon, peccary, and a primitive pronghorn antelope.

BUILDING STONE QUARRIES

Many of the fossils used by the early British geologists to establish geologic horizons were collected in limestone and sandstone quarries in England, Scotland, and Wales. Many of these are still being worked today, and the fossils keep showing up. In September 1981 I visited one of these quarries in southern England as the fossils were everywhere.

Like pit operators and earthmoving contractors, quarry operators may give you permission to prospect the quarry. In all of these operations there will probably be some fossils in or around the office building if there are any in the excavation.

In Arizona, near Seligman, they quarry the so-called Kiabab Stone, actually the Coconino Sandstone of Permian age. These slabs of fine-grained sandstone are often covered with small tracks and trackways of amphibians and reptiles. The stoneyard operators often separate track slabs from the rest and sell them to interested collectors.

COAL MINES

Coal is formed when great masses of plant material forming in coastal or inland swamps is compressed by the weight of overlying sediments. Often imprints of leaves and tree trunks are preserved in the coal or in the interbedded sediments.

Sifting through the dumps of underground mines should produce a variety of fossils ranging from leaf imprints to dinosaur footprints. Near Bernissart in Belgium a coal mine discovery in 1877 yielded 23 skeletons of the dinosaur *Iguanodon*.

Strip mines present an even better opportunity for fossil collecting as so much more rock is removed to expose the coal and it is more readily accessible. One of the most famous strip-mining areas, fos-

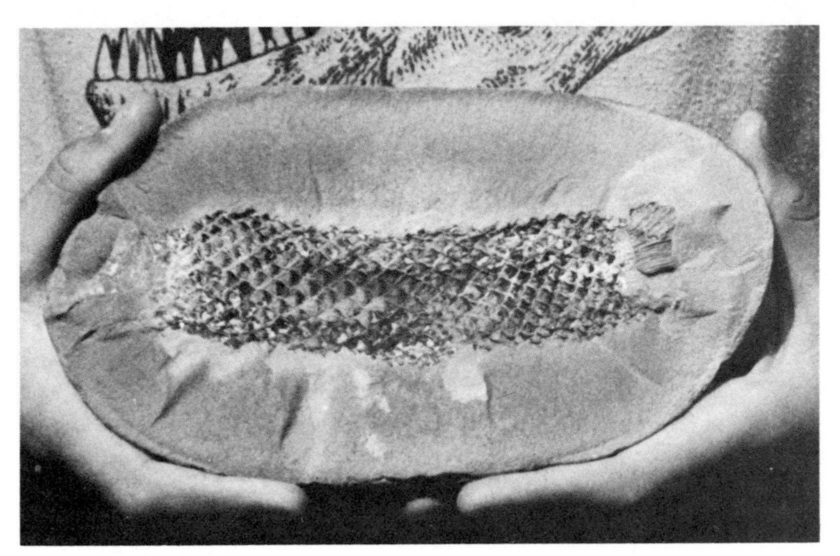

Figure 9. Lepidodendron, *a Pennsylvanian tree trunk coalified in a nodule.*

silifcrously speaking, is the Mazon Creek area south of Chicago where thousands of fossils have been found in brown ironstone concretions. These concretions contain leaves, jellyfish, chitons, insects, spiders, worms, larval fish, amphibians, and strange creatures like the mysterious Tully Monster which still defies classification.

The middle Pennsylvanian Francis Creek Shale contains tons of these concretions as it is being removed to uncover the underlying coal bed. The Peabody Coal Company's Pit 11 in Will and Kankakee counties is a major fossil site.

CONSTRUCTION SITES

Wherever large quantities of sedimentary rock are being moved there is a chance of finding fossils. One problem is that today's earthmoving equipment is so large that the operator rarely sees fossils when they are uncovered. Years ago I visited an aqueduct ditching project east of Berkeley, California. While surveying for final grade, the stake artist had hit a mastodon tusk; in the side of the cut nearby were the lower jaws with the back end sheared away. A search of the spoil along the ditch showed that what had probably been a whole skeleton had been chewed up and spit out by the ditcher without the operator seeing a bit of it.

The same problems of access may be encountered here as at sand and gravel pits. If you can't get on the site, then find out where the spoil is being dumped and look there.

Figure 10. Construction sites often produce fossils, but the size of the machinery generally interferes with collecting.

Figure 11. A Cretaceous cycad stump from the construction site in the Black Hills shown in Figure 10.

When I was with the Los Angeles County Natural History Museum we were often invited to construction sites where fossils were found. A storm sewer excavation passed through a small buried tar pit which yielded a variety of late Pleistocene mammals. A major light industry complex being built on very fossiliferous Miocene marine shales and silts was opened to us on weekends. I took the local fossil club to the site on several weekends where we made a fine collection of fish, shore and sea birds, and marine mammals.

Chapter Three

SOMETHING ABOUT GEOLOGY

TOPOGRAPHIC MAPS

*M*aps are not mysterious documents designed to conceal information and strike terror in the heart of the beholder. Anyone who has sketched directions on a scrap of paper or scratched them in the dirt has drawn a map; anyone who has used such a sketch and arrived at his destination has successfully read a map.

Maps are graphic representations of a part of the earth's surface and are drawn to a scale so that true distances between points can be measured. Maps are often colored to help the reader differentiate between different sorts of natural and cultural features. Man-made objects and most natural features are generally represented by little symbolic drawings that suggest the nature of the feature. Some of these conventional symbols are shown on the U.S. Geological Survey's Topographic Map Symbols sheet. In order to indicate elevations and the shape of the earth's surface, many maps have contours which are imaginary lines of equal elevation drawn at a fixed vertical interval on the earth's surface above or below a designated reference plane. In addition, most maps are oriented so that north is toward the top of the sheet.

MAP SCALE

The scale of a map is the ratio of the map's size to the earth's size and is often expressed as a representative fraction like $^{1}/_{10,000}$ (or 1:10,000), which means that 1 unit of measurement on the map equals 10,000 equivalent units of measurement on the ground. These units of measurement could be millimeters, centimeters, inches, or thumb-widths, whichever unit you happen to prefer at a given time. What is critical in measuring distances on a map is *not* what unit you choose, but for you to know how many units on the ground equal one of that same unit on the map. Sometimes map scale is written as a statement—for example: "1 inch equals 25 miles." This simply means that 1 inch on the map equals 25 miles on the ground.

The U.S. Geological Survey is the principal federal agency responsible for preparing, manufacturing, and distributing the topographic maps that cover the entire surface of the United States. The Survey most commonly publishes 7½-minute (7½′) and 15-minute (15′) *quadrangle* maps at scales of 1:24,000 and 1:62,500, respectively. The quadrangle maps (or "quads") are bounded at their northern and southern edges by lines of latitude and at the eastern and western sides by lines of longitude spaced at intervals of either 7½ or 15 minutes of circular arc on our globe. A minute of arc equals $^1/_{60}$th of 1 degree of arc or $^1/_{60} \times {}^1/_{360} = {}^1/_{21,600}$ of a circle.

THE LAND SURVEY SYSTEM

Many states have been divided into land divisions based on an east–west parallel of latitude called a *base line* and a perpendicular north–south meridian of longitude known as the *principal meridian*. In central California the Mount Diablo base line and principal meridian are the basis for many of the land survey divisions in the Far West. Tied to these lines are 6-mile-wide squares known as *townships*. These 36-square-mile areas are numbered north and south from the base line as *tiers*, but usually called townships, and east or west of the principal meridian as *ranges* (see Figure 12).

The 36-square-mile townships are further divided into 1-mile squares known as *sections*. Sections are numbered from 1 to 36 beginning in the northeast corner of the township as shown in Figure 12. The sections are further divided into halves or quarters and each half or quarter again into halves or quarters, and so on. The black block in the section diagram is designated as the NW-¼ of the NE-¼ of Section 22, T2S, R3E.

The reading of a topographic map is simple and straightforward. Be sure you know the scale and the contour interval when you begin to study any map.

As their name implies, topographic maps depict the topography or shape of the earth's surface. The commonest way of showing elevations and the shape of the land is with *contour lines*, which connect all points of equal elevation on the earth's surface above or below a selected datum plane. For U.S. Geological Survey quadrangle maps the reference plane chosen is mean sea level. The vertical distance or elevation difference between adjacent contour lines is called the *contour interval* and is selected by the cartographer (map maker) according to the *relief* of the terrain to be represented. (Topographic relief is the difference in elevation between the highest and lowest points in a region.) In steep terrain such as along the rugged crest of the Sierra Nevada a

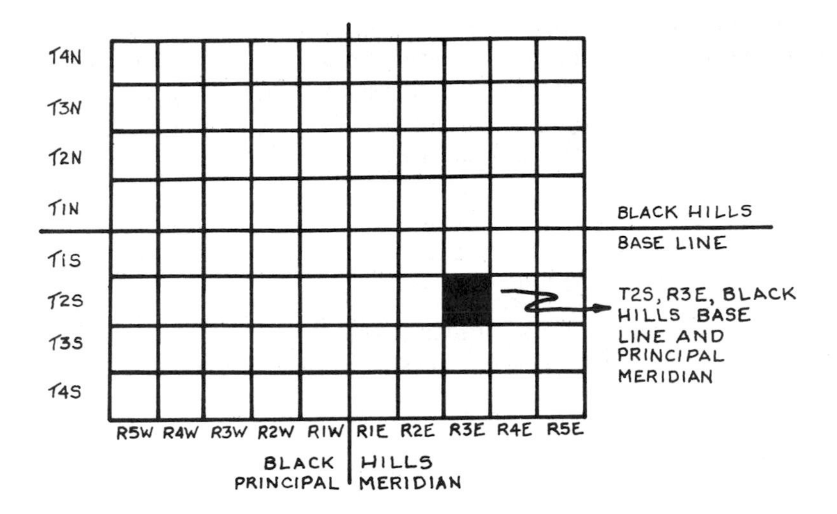

R3E

T2S

6	5	4	3	2	1
7	8	9	10	11	12
18	17	16	15	14	13
19	20	21	22	23	24
30	29	28	27	26	25
31	32	33	34	35	36

SECTION 22, T2S, R3E, BLACK HILLS BASE LINE AND PRINCIPAL MERIDIAN

NW¼

22

S½

NW ¼, NE ¼, SECTION 22, T2S, R3E, BLACK HILLS BASE LINE AND PRINCIPAL MERIDIAN

Figure 12. The Land Survey location system.

contour interval of 50′ (50 minutes) or even 100′ might be appropriate, but in the center of California's Great Valley a contour interval of 10′ is needed to show the configuration of the land. Where the contours are close together on a map, the land surface they depict is steep; where they are far apart, contours indicate a gentle surface. Depressions without outlets (closed depressions) are shown by *depression contours* whose short bars point into the depression.

Maps published by government agencies like the U.S. Geological Survey generally use the following color conventions:

- Black and red: Works of man (roads, building, etc.) including boundaries and land survey lines
- Pink: Urban area
- Blue: Water
- Green: Vegetation (except grass)
- Brown: Topographic and geologic features

MEASURING DISTANCE ON A MAP

With the kind of scale information provided by Figure 13 it is very easy to measure the straight-line distance between any two points on a map.

1. Imagine that the line A–B in Figure 13 is drawn on a 7½′ quad map. With a piece of paper or a ruler determine the map distance from point A to point B. Lay this distance out on the scales in Figure 13 and find the equivalent or true distance on the ground in miles, feet, and kilometers.

2. Measuring distances along a curved line like a meandering stream or a twisting mountain road is a more complicated process. You may use a map measurer, a small device with a wheel you push along the route to be measured. As the wheel turns, it moves an arrow on a dial that indicates the distance traveled. Or you may divide the curved line to be measured into a series of nearly straight-line segments and then mark the length of each segment on the edge of a piece of paper so that the total accumulated distance can then be measured using the map scale.

THE STEREOSCOPE

Humans have what is known as stereoscopic vision—in other words, we see the world around us in three dimensions (3-D). Our eyes are mounted on the front of our heads with their pupils or centers approx-

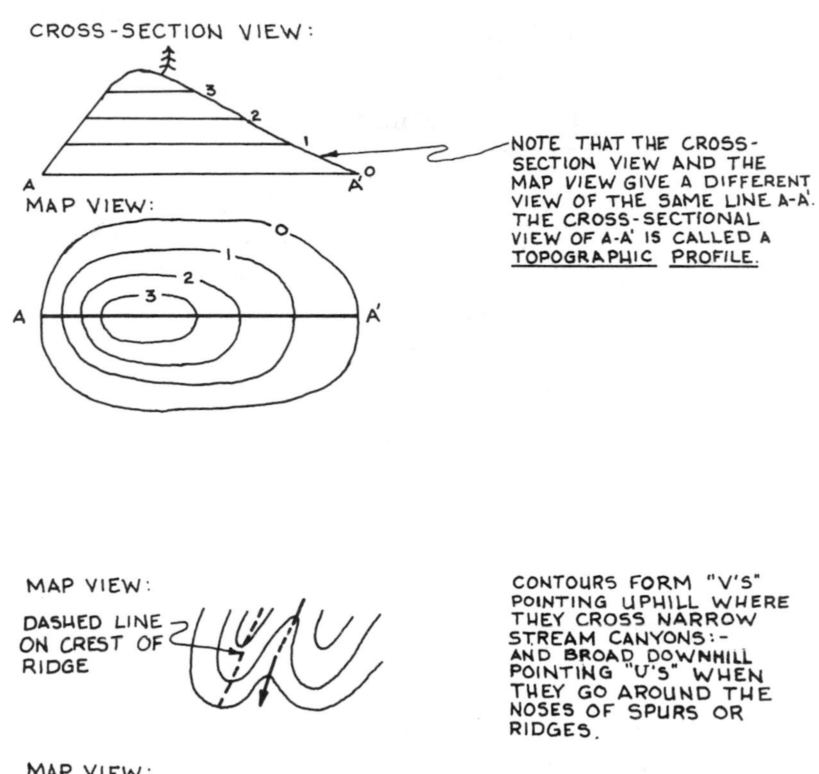

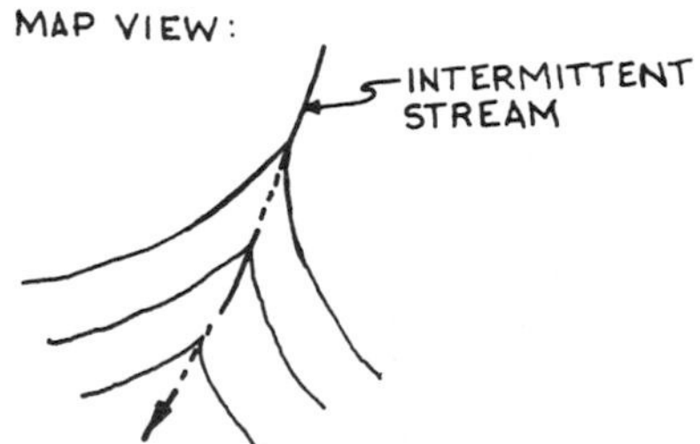

Figure 13. How contour lines show the shape of the land.

imately 2.5 inches apart. We see two different images when we look at any object. These images are merged by our brain to give us a 3-D picture with depth. The same effect can be conveyed by overlapping photographs when viewed individually by each eye. The parlor stereoscope of the past was such a device, as are the numerous viewers sold in toy stores and curio shops.

Aerial photographs may be taken in overlapping sequences so that they can give us a stereoscopic view of the ground.

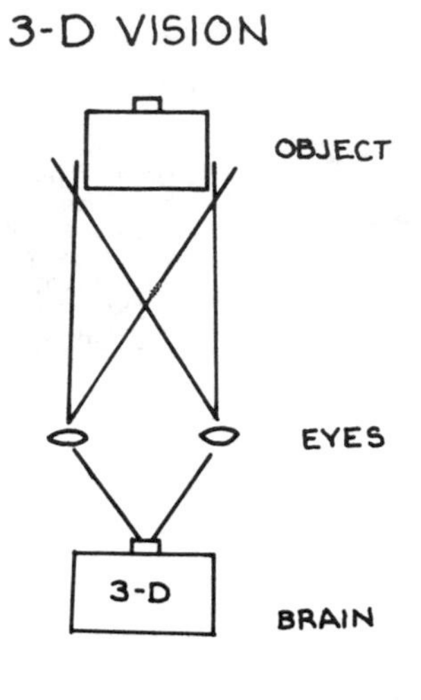

A.) HOW WE DO IT WITH OUR EYES AND BRAIN.

B.) TAKING OVERLAPPING AIR PHOTOGRAPHS.

C.) CONVERTING OVERLAPPING AIR PHOTOS TO A 3-D IMAGE.

Figure 14. Three-dimensional vision and how the stereoscope works.

Two types of stereoscopes to view such photographs are commonly used. The pocket stereoscope is a small folding device which is relatively inexpensive and ideally suited for use in the field. Two lenses are mounted in a frame which may be adjusted for interpupillary distance. The frame is held above the table and photographs on folding legs. When the photographs are adjusted so the overlapping portions are viewed separately, a stereoscopic image is formed by the brain. Much more expensive and better instruments are the mirror stereoscopes which have the disadvantage of being bulky and hard to handle in the field. The body of the instrument carries the eyepieces with two prisms at their base which receive the image as a reflection transmitted by two mirrors mounted at the ends of the frame. The frame is supported above the table by folding legs, one of which is adjustable for length. Care must be taken with the mirrors as they are "first-surface" mirrors with the reflective coating on the outside where fingerprints may cause corrosion. Magnifiers are available to cover the eyepieces and greatly enlarge the picture while reducing the area being viewed.

Set up the stereoscope. If the lenses or mirrors are dirty, carefully clean them with tissue and alcohol. Take two adjoining photos and orient them so that the serial number is in the upper left-hand corner. Check the overlap so that the left side of the left photo doesn't match any part of the right photo and that the right side of the right photo doesn't match any part of the left photo. Place the photos side by side and then select a prominent spot near the center bottom edge on each photo. Place an index finger on the spots and then, looking through the stereoscope, move the photos until your fingers overlap. The photos will now be approximately arranged for stereoscopic viewing. Carefully adjust the photos until you get a full, nonfuzzy 3-D image.

If you have long hair you will want to pin it at the back of your neck or on your head so that it doesn't hang down on the stereoscope.

GEOLOGIC TIME

As soon as we begin dealing with the relative positions of rocks and their relations to each other we have to introduce the element of time into our geologic thinking.

Geologic time may be thought of in terms of *real* or *absolute time* where we are actually dating events by radiometric means and citing the events in terms of millions of years before the present (m.y.b.p.). Since radiometric dating is becoming more and more reliable, this method of dating is now widely used in geology.

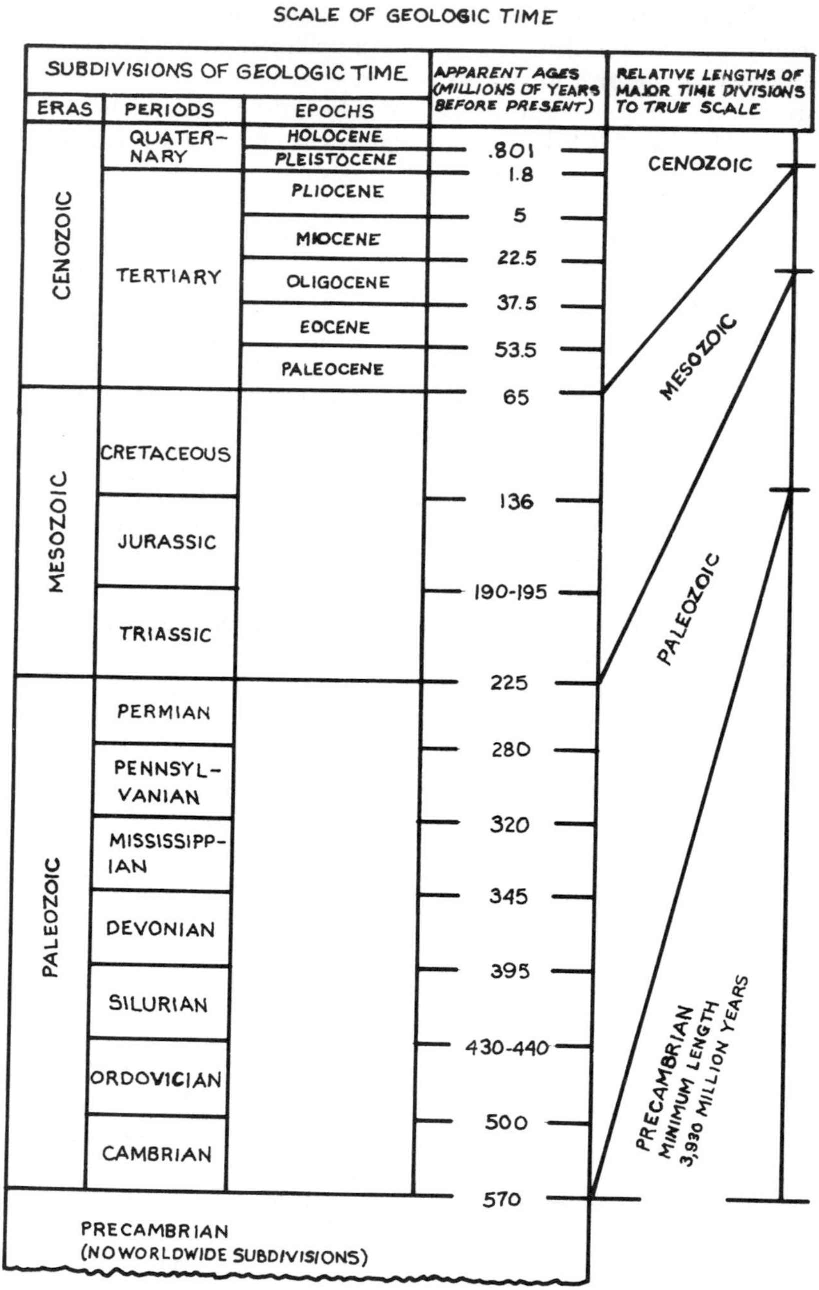

SCALE OF GEOLOGIC TIME

SUBDIVISIONS OF GEOLOGIC TIME			APPARENT AGES (MILLIONS OF YEARS BEFORE PRESENT)
ERAS	PERIODS	EPOCHS	
CENOZOIC	QUATERNARY	HOLOCENE	.801
		PLEISTOCENE	1.8
	TERTIARY	PLIOCENE	5
		MIOCENE	22.5
		OLIGOCENE	37.5
		EOCENE	53.5
		PALEOCENE	65
MESOZOIC	CRETACEOUS		136
	JURASSIC		190-195
	TRIASSIC		225
PALEOZOIC	PERMIAN		280
	PENNSYLVANIAN		320
	MISSISSIPPIAN		345
	DEVONIAN		395
	SILURIAN		430-440
	ORDOVICIAN		500
	CAMBRIAN		570
PRECAMBRIAN (NO WORLDWIDE SUBDIVISIONS)			

RELATIVE LENGTHS OF MAJOR TIME DIVISIONS TO TRUE SCALE

CENOZOIC

MESOZOIC

PALEOZOIC

PRECAMBRIAN
MINIMUM LENGTH
3,930 MILLION YEARS

Figure 15. The geologic time scale.

Most geologic dating however, is still done in terms of *relative time* (younger events vs. older events) and is based on the *geologic time scale,* a series of arbitrary divisions based on breaks in the geologic record of Western Europe, developed by early European, mainly British, geologists. The scale was designed to show relative time and the temporal position of rocks anywhere in the world.

The *period* is the basic time unit based on a designated rock sequence called a *system.* Periods are grouped into *eras* and subdivided into *epochs,* which in turn are often broken down into *ages.*

When the geologic time scale was established, there was no way of determining the actual time involved or the length of the subdivisions. As the named units were based on local geologic events, the length of these divisions are all unequal, which contrasts with our normal practice of breaking time up into units of equal length such as seconds, minutes, and hours.

The zero point in this scale is the base of the fossiliferous rocks exposed in Wales. These sediments and their contained fossils of primitive marine life were deposited on much older distorted rocks which had been laid down before life had begun to leave much of a fossil record. This earliest period is called the Cambrian Period, from Cambria, the ancient Roman name for Wales. This is the first period of the Paleozoic Era ("time of the ancient life"), which in North America is divided into six periods because we have split one of the European periods, the Carboniferous, into the Mississippian and the Pennsylvanian. As European rock sequences were studied and placed into their correct position in the sequence of geologic history by relative age dating techniques, the time scale was slowly developed.

Relative age dating techniques used by geologists to establish the geologic history of an area include:

1. Superposition—younger layers are deposited on top of older layers.
2. Cross-cutting relationships:
   a. Faults—faults are younger than rocks they offset but older than rocks they do not displace.
   b. Intrusions—intrusions are always younger than the rocks they intrude.
3. Inclusions—An inclusion is always older than the rock that includes it. The pebble in a conglomerate is always older than the layer itself.
4. Fossils—When it was discovered that fossil faunas and floras

or ancient living communities evolved and changed through time, it was possible to date sediments by their fossil content in a relative manner.

GEOLOGIC MAPS

A basic tool of the geologist is a map showing the outcrop pattern of rock units in a certain area. An *outcrop* in geology is the area where a certain sort of rock is exposed on the surface or would be if there were no covering of soil or post-Tertiary sediments. Some maps which show surficial deposits will include Pleistocene and Holocene deposits.

Geologic maps are constructed on either topographic maps or air photos when the latter are available. In areas where neither are available any map may be used as a base map.

The *formation* is the usual rock unit used in geologic mapping. This is defined as a rock or lithic unit that is extensive enough to be shown on a map and has contacts that can be used to distinguish it from adjacent formations. By implication the map should have a scale no greater than 1 inch to the mile.

Investigators select convenient rock units for mapping purposes. These units are given geographic names. If the unit is made up of several types of rock, it is referred to as a formation—for example, the Supai Formation. If the unit has a predominate rock type, it is named for that rock type—for example, the Redwall Limestone, the Coconino Sandstone, or the Hermit Shale.

On the map each formation is color-coded by age and given a letter symbol indicating the age and the formation's name:

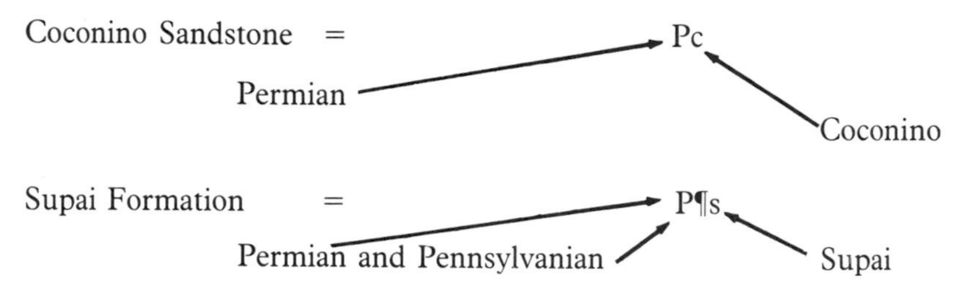

The attitude of the beds is shown by strike and dip symbols which show the direction the bed trends across the surface and the direction and angle at which it tilts into the ground. These and other symbols used to indicate rock attitudes are discussed below.

Geologic maps are primarily published by the United States Geologic Survey and the various state geologic surveys or bureaus of mines.

Symbol and Color Code Chart

Symbol	*Period*	*Color Used for Sedimentary Rocks*
Q	Quaternary	Light yellow and gray
T	Tertiary	Yellow and orange
K	Cretaceous	Green and yellow green
J	Jurassic	Light blue green
Tr	Triassic	Blue green
P	Permian	Blue
¶	Pennsylvanian	Pale turquoise
M	Mississippian	Turquoise blue
D	Devonian	Purple
S	Silurian	Violet
O	Ordovician	Pink
Ꞓ	Cambrian	Salmon
prꞒ	Precambrian	Medium browns for younger part; violet-gray and dark salmon for the older part

THE ATTITUDES OF ROCK STRATA, DIKES, SILLS, AND FAULT PLANES

In order to record the geologic data of a fossil locality, the attitude of the strata should be noted. If the beds are not horizontal, if they have been faulted, intruded by dikes, or sills, the strike and dip of these features should be noted and entered on your locality map or air photo.

Rock strata, dikes, sills, and fault planes are all planes which may intersect the earth's surface. The trace of this intersection, the exposed edge of the plane, as it runs across the countryside can tell the trained observer a great deal about the attitude of the plane (its orientation in space) and the character of its contact with adjacent rocks.

Horizontal planes will follow contour lines around hills and valley walls.

Vertical planes will form straight lines across the earth's surface as the trace is followed over hills and across canyons.

Planes with a tilt (dip) between horizontal (0°) and vertical (90°)

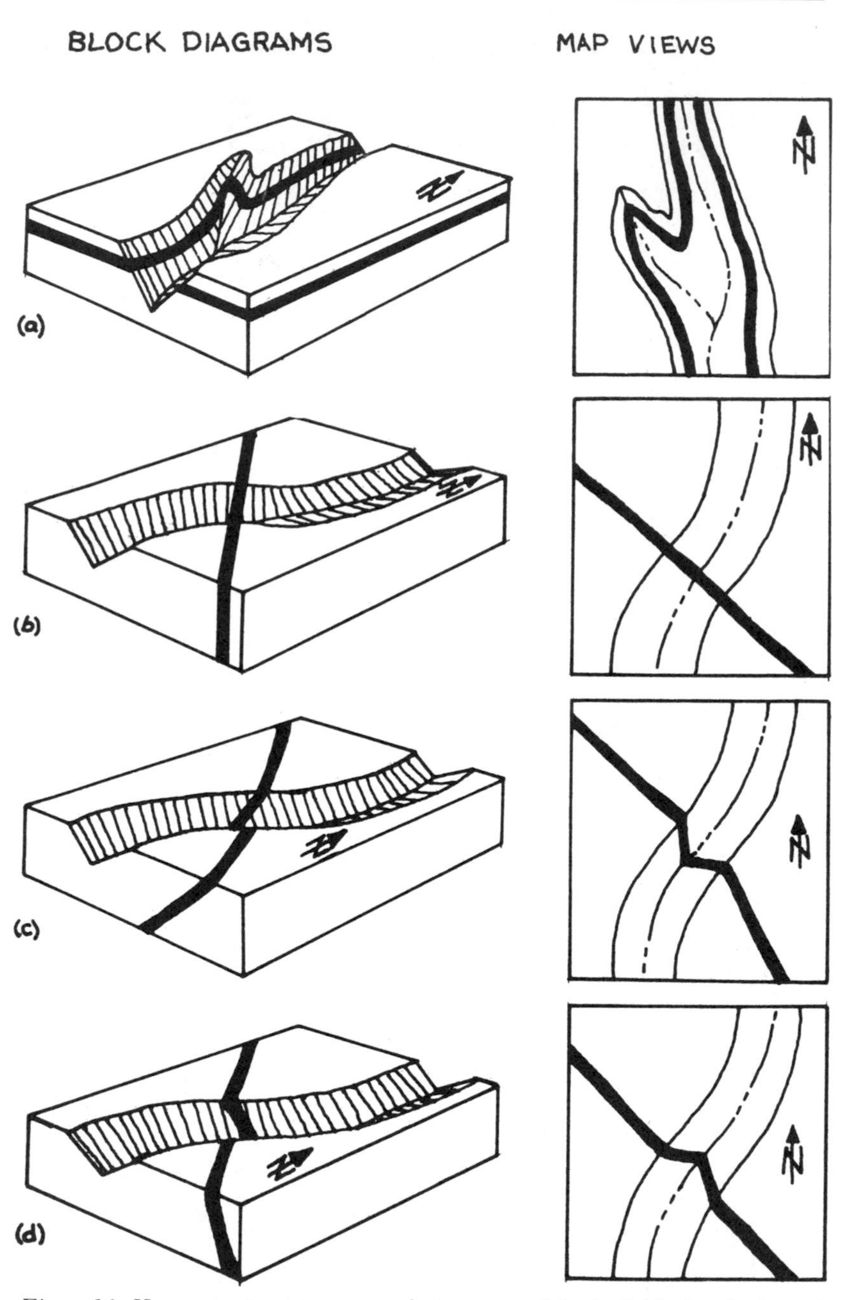

Figure 16. How outcrops appear on geologic maps and in the field when horizontal (a), vertical (b), and tilted or dipping (c and d).

will form "V's" where they cross canyons or valleys with the "V" generally pointing in the direction of dip. The exception to this rule is that the trace will form a "V" that points upstream when the plane is dipping downstream at an angle that is less than the gradient of the stream. If the dip is the same as the gradient of the stream, the trace will not cross the stream but will parallel the stream's course.

Remember: The trace of a plane surface has nothing to do with contour lines. For example, contours lines form "V's" that always point upstream, but plane contacts "V" up- or downstream depending on the direction and amount of dip of the plane.

Once the direction and amount of inclination of the contact has been measured by a geologist, this information can be shown on a map with the following symbols:

⊕ = Horizontal layers
⋌32 = Inclined layers
\+ = Vertical layers

These symbols are commonly called *strike* and *dip* symbols. Recall from plane geometry that two intersecting lines will determine a plane. Geologists also mentally draw two intersecting lines on the geologic plane they wish to depict on a map; one line—the dip line—is drawn down the steepest direction of the plane, the direction water would flow down the plane. The other line—the strike line—is a horizontal line drawn on the plane's surface and thus is perpendicular to the dip or

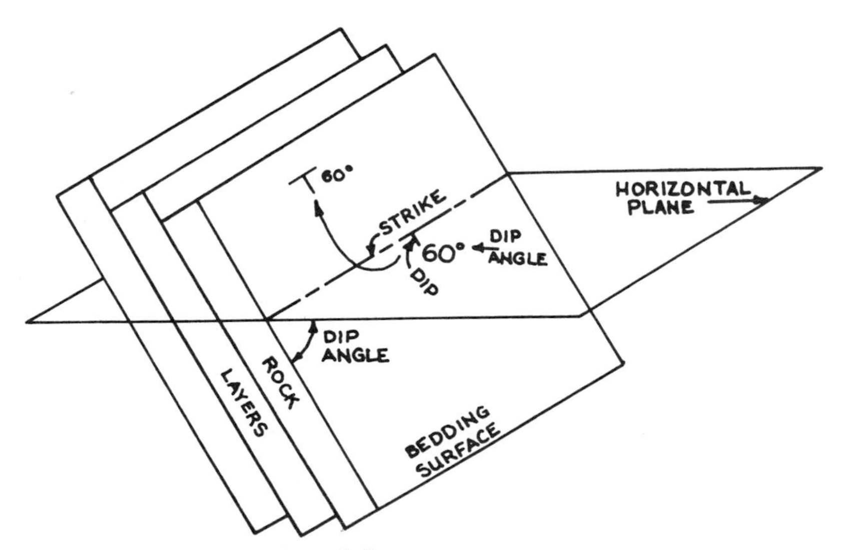

Figure 17. Attitude or strike and dip.

steepest line on the plane. See Figure 17. Strike is defined as the compass direction as measured from *north* of the level line drawn on the plane. Dip is normally recorded as the number of degrees that the plane is inclined downward from the horizontal. A horizontal plane has neither a strike nor a dip, and the steepest dip a plane can have is vertical (90°). To show this information on a map a long line is drawn parallel to the strike of a layer and a short line is drawn perpendicular to it in the direction of the dip. The symbol is placed on the spot depicted on the map where the attitude was measured in the field.

Chapter Four

COLLECTION, PREPARATION, AND CURATION

While discussing collecting and preparation, I'll refer mainly to vertebrates because they are my own specialty and generally are more difficult to collect and prepare than either plants or invertebrates. So let's get on with it and assemble our equipment to go into the field.

Selecting field equipment is very much like choosing a spouse. It's all a matter of individual taste. I'll tell you what I prefer, what some of "them other people" take, then you can pick out what you like. Years ago I got into a series of arguments with Morris Skinner of the Frick Laboratory, American Museum of Natural History, about field techniques. I hate to admit it, but I consider Morris the finest fossil collector in the profession. However, he does have some very peculiar techniques and does things I wouldn't allow any of my people to do. The upshot of this two- or three-day shouting match was that we both agreed that the other fellow did get his specimens home in one piece, so that although his methods were completely without merit, they did seem to work.

This brings us to the two basic rules of fossil collecting:

1. Get the specimen home in the best possible shape.
2. Make it as easy as you can on the fellow in the laboratory.

Another rule is perhaps even more important: *Get all the geographic and geologic data you can on the specimen so that it will be scientifically important and not simply a trophy or a curiosity.*

I divide my field equipment into two parts: (1) the things that stay in the truck until needed, and (2) the items I carry with me. The truck list includes:

- One or more 5-gallon cans of water
- (U.S. Army) Surplus ammunition chest filled with 35–50 pounds of plaster
- Burlap sack full of burlap sacks
- A dozen or so old newspapers (*New York Times* or *Los Angeles Times* size) or several rolls of toilet paper

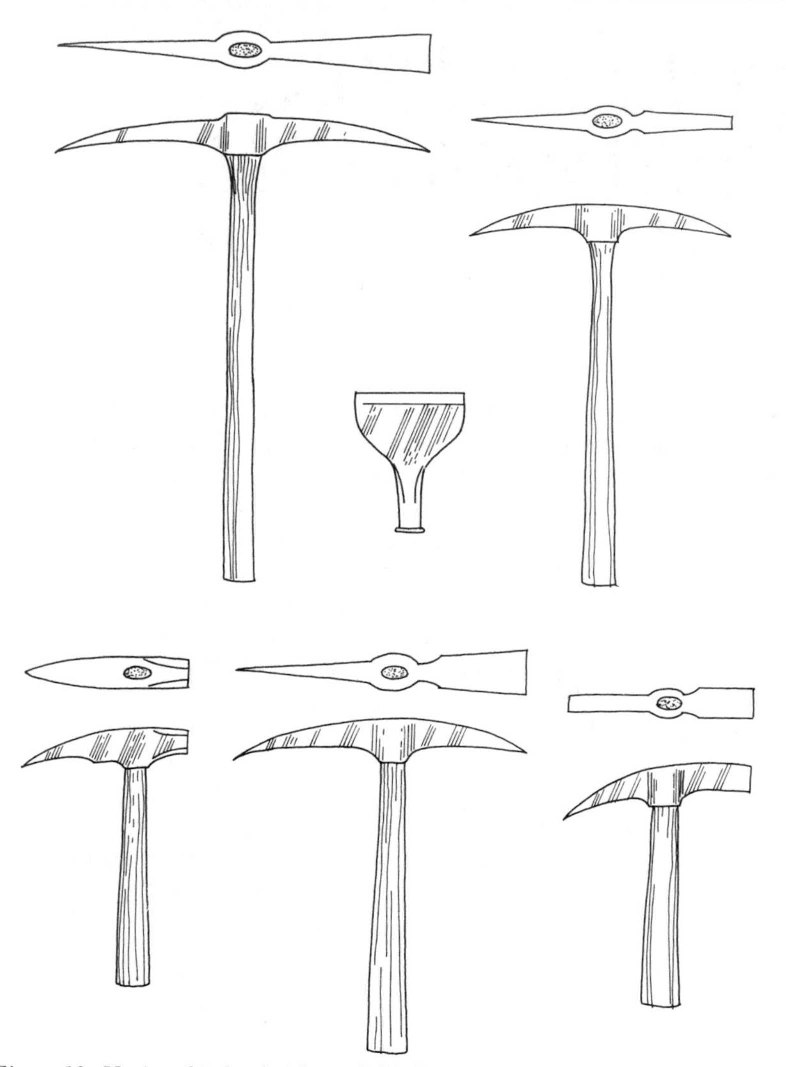

Figure 18. Various kinds of picks and chisels.

- A gallon of shellac
- 1 or 2 gallons of alcohol or of Glyptal thinned with acetone
- A gallon of white glue
- Plaster-mixing bowl (plastic wash basin or a gold pan)
- Axe or hatchet
- Saw
- Hammer
- Square

- Bag or box of 8-penny nails
- Railroad pick
- Long-handled shovel
- Pair of fence pliers
- Bag of fence staples
- Insect repellant
- Geologist's hand pick or hammer (a heavy one)

The fence pliers and staples aren't always necessary, but sometimes, *with the owner's permission,* you will find them handy for dismantling and repairing a stretch of fence when there's no nearby gate for your truck or car.

The personal list includes:

- Surplus GI pistol belt
- Surplus GI canteen and cover, hung on belt
- Surplus GI first-aid packet pouch, containing a throat lozenge can and a packet of cigarette papers
- Surplus gas mask bag (the side-carrying type with a waist strap). I use this for a field bag, as it has side pockets for small things and carries my tools without any space left over for bulky, heavy fossils.
- Marsh pick. I carry this in preference to a hand pick as it is helpful in climbing. Instead of the regulation handle I make my own from a cutdown double-bitted axe handle. Serious digging will require other things from the truck, so I can get a hand pick at that time. My oldest son has a theorem which states: Anyone who carries a Marsh pick will borrow my hand pick to dig with.
- A pusher (6-inch piece of ¼-inch hex rod flattened and sharpened at one end and on the sides of curve)
- Plastic shellac bottle with a brush fitted in the tip
- Whisk broom
- A couple of 1- or 2-inch paintbrushes
- Heavy jackknife or hooknosed knife
- Snake bite kit. One day in southwestern South Dakota a geologist friend on one of my trips stepped on a rattlesnake. He freaked out. I thought it was just about the funniest thing I'd ever seen. The next day I stepped on one. I was 10 feet in the air and running when I was ready for a coronary!
- A few sheets of newspaper
- A few sheets of toilet paper
- Small roll of masking tape
- Felt-tip marking pen
- Field notebook

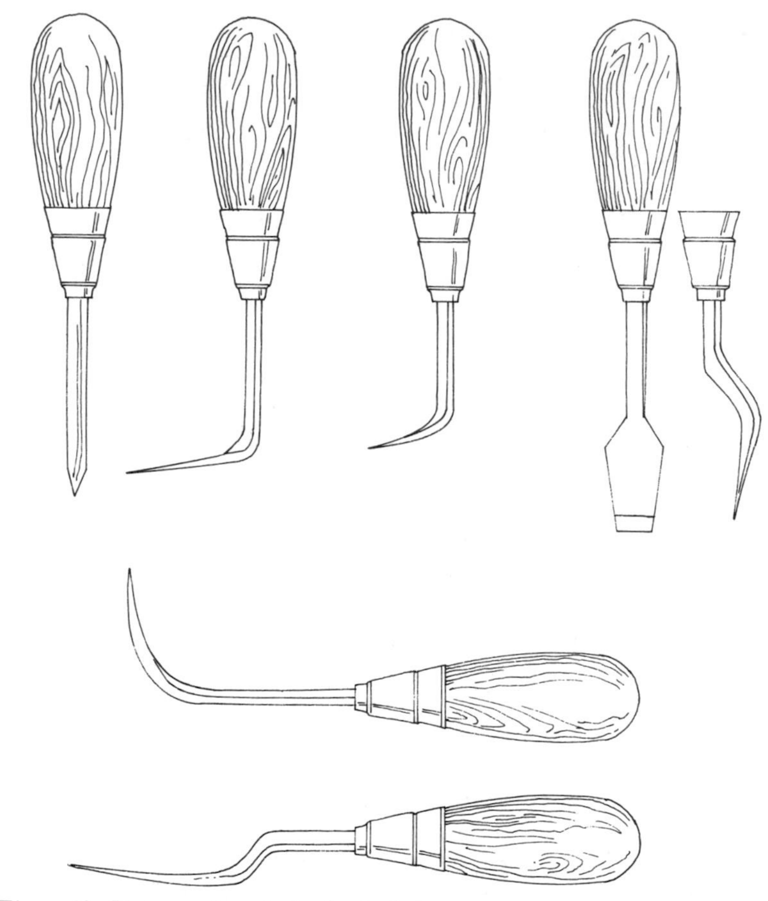

Figure 19. Home-made awls for fossil digging.

With this array I feel adequately supplied to go out and do a job of fossil prospecting.

As far as "them other people" are concerned, some carry large knapsacks as collecting bags. They also prefer a supply of paper bags to old newspapers for wrapping small specimens. I believe that a specimen tightly wrapped in newspaper, the same way your butcher wraps a piece of meat, makes a more protective package than a paper bag. Others prefer Glyptal to shellac for a hardener.

Some prefer a pre–World War II entrenching pick to either a Marsh pick or geologist's hammer, and you will always see a number of ice picks for digging in any party. All these things are useful; it's just a matter of individual preference. For that matter, many collectors go into the field with hardly more than an ice pick and a few sheets of toilet paper. This usually means returning to the truck whenever you find something, and I don't recommend it.

Fossil prospecting and finding is a matter of self-training and experience. There is no royal road to discovery; you just have to get out into the areas of bare rock outcrops and begin a slow and careful search of the ground for fossils being exposed by weathering. The steppes and deserts of the world are the best places, because there the ground is not covered with weathered soil.

Tiny jaws and teeth may be easily spotted when your eye becomes trained to the task. A good prospector should be able to spot a rabbit jaw at 10 feet. If you find small material, you should begin a hands-and-knees search of the area, which may be especially rewarding. If the rock appears to be rich in small fossils or bone scraps, gather a sackload of it for washing, screening, and hand-sorting. Matrix which won't break down in water can be soaked in buckets of kerosene for 24 hours or so. Then pour off the kerosene, refill the bucket with water, and put it on the stove to boil. Usually the most stubborn matrix can be broken down this way. Care should be taken as the kerosene being pushed out of the rock by the water and boiled off will rise to the surface and may ignite. You might be more popular with your spouse if you performed this operation in the backyard on your camp stove.

Small fossils can be wrapped in cigarette paper or tiolet tissue and stored in the throat lozenge can. Larger fossils, which may be picked up without further treatment beyond a possible shellacking to seal cracks and hold small broken pieces together, should be wrapped in newspaper. The locality should be marked on all packages with a felt-tip pen.

When you find leads of bone on a weathered surface or slope, stand back and survey the situation. Gather up all the loose pieces that have washed away from the main fossil. Remember that patience is the main virtue in fossil collecting. Most vertebrates should be left for experts to collect. They are rare, often fragile, and a few minutes of bungling by an overenthusiastic, untrained collector can destroy a scientifically valuable find. If the specimen is soft, cracked, or run through with fine roots, dowse it with one part white shellac thinned with about four parts alcohol.

Do not try to work with any part of the fossil while the shellac is wet; let it get really dry and hard before touching it again. While it dries, you can begin to dig carefully around the specimen with your pusher and awls. Keep the work area clean with your brushes and whisk broom, so you can see what you're doing. A major rule when bringing home the fossil is: When in doubt, plaster.

At the extremes of plaster block making, I have worked with one block which contained a complete brontothere skeleton, less the skull which was in another block. This one weighed 2,600 pounds. The local National Guard brought a tank recovery vehicle into the Badlands to load it onto our truck. The smallest I ever made was for a 2-inch Oligocene "deer" skull. I used a torn-up pocket handkerchief for that one. I wasn't married, and I was allowed to tear up my pocket handkerchiefs in those days.

Figure 20 a. A skeleton as it might appear when discovered.

Figure b. The skeleton has been exposed and trenched. There is actually too much of the skeleton exposed in this picture. A better pad of rock should surround the bones.

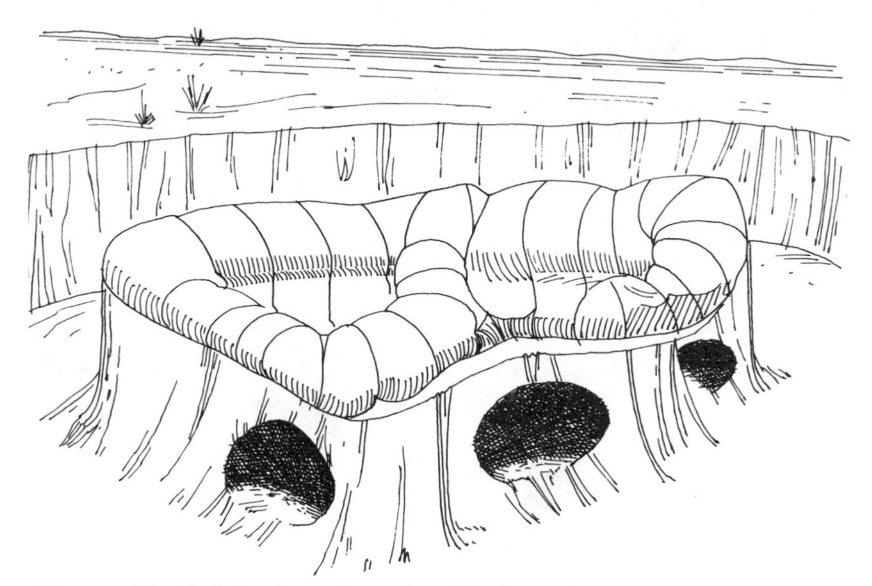

Figure c. The block has been plastered and further undercut.

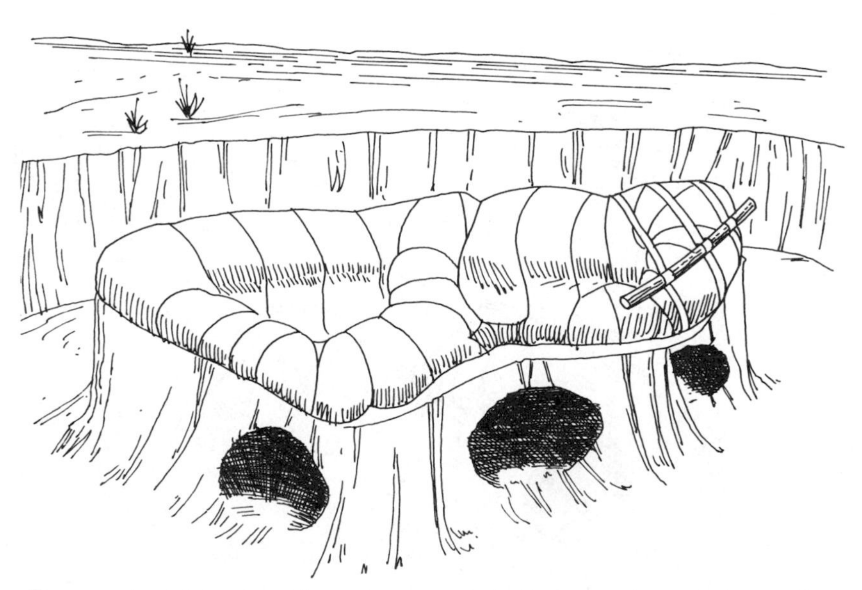

Figure d. A stick has been plastered to part of the block to make a carrying handle.

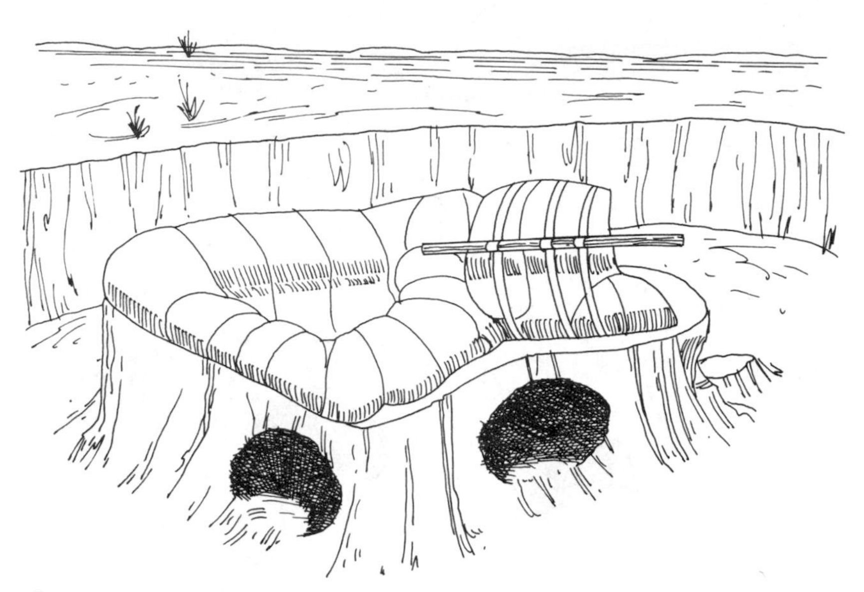

Figure e. One block has been removed and another carrying stick plastered in place.

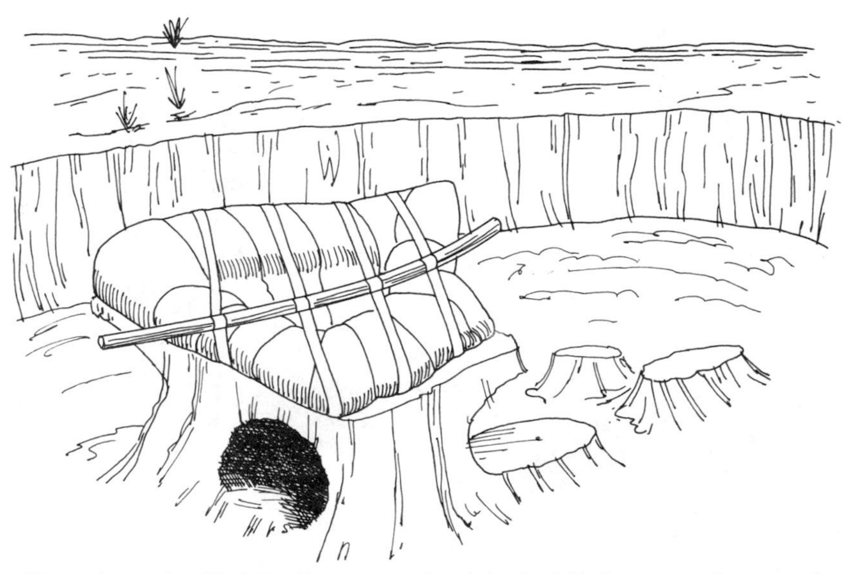

Figure f. Another block has been removed and the final block prepared for removal.

To prepare for plastering, first dig a trench completely around the specimen to several inches below its suspected bottom. Undercut the fossil and trim the matrix so that there is an inch or so of rock all around. If the specimen is large and the matrix is soft, you will want to dig tunnels under it so cinches of plaster-soaked burlap can be used to tie the block together.

Cut burlap into strips the length of the bags and 3 or 4 inches wide. I pull out the long strings of jute thread, which keeps me from having to fight them when I am plastering. Roll the strips like bandages and put them to soak in the basin of water. Soaked burlap will not pull the water out of the plaster and upset its setting and proper hardening. When the strips are ready and wrung out, and the shellac on the specimen is thoroughly dried, wet down the matrix so it won't pull water out of the plaster and so the plaster will bond to the matrix. Cover any exposed bone with wet newspaper or wadded wet toilet paper to protect it from plaster. Now you're ready to start.

Mix the plaster (I use Red Top or Gold Bond molding plaster or the equivalent, which can be bought in 100-pound sacks in most lumber yards) to the consistency of whipped cream. Run burlap strips through the plaster one at a time, making sure they are soaked with it, and lay them on the block. Smooth each one down firmly and push it gently into the depressions so that there will be no movement of the block

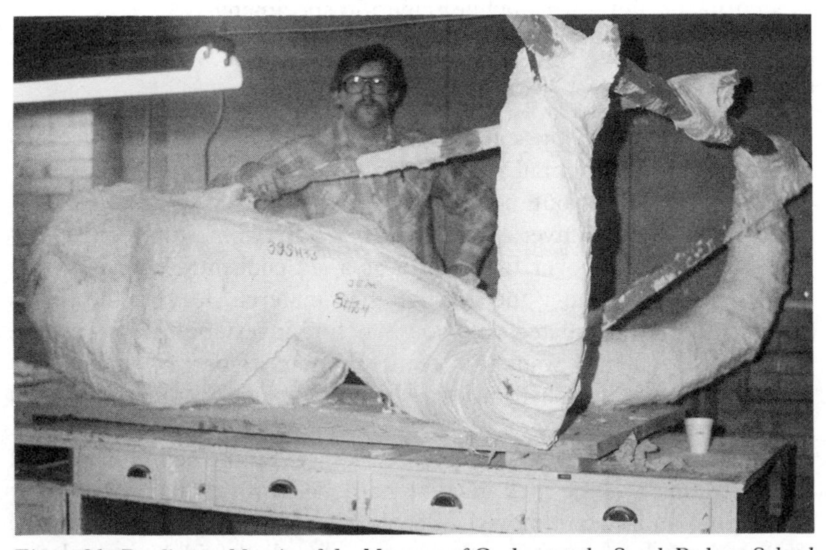

Figure 21. Dr. James Martin of the Museum of Geology at the South Dakota School of Mines and Technology with a plastered mammoth cranium from a paleo-Indian kill site in southwestern South Dakota.

beneath it. Excess plaster can be removed by pulling the strips between your fingers before you lay them on. Cover the block with at least two layers of overlapping burlap. I lay a course one way with the ends tucked into the undercut, overlapping the strips by about an inch. I then repeat the process at right angles to the first layer. Finally, if this is enough, I'll run one or more long bands around the bottom of the specimen and down in the undercut to bind all the ends together. If the block needs bracing you can plaster in pieces of boards or sticks. On the big brontothere block we used steel fence posts.

When the plaster has set, usually in 30 to 60 minutes, you can continue undercutting until the block breaks loose from its pedestal. With a great deal of caution turn over the block. If it is improperly made, the rock may drop out of the jacket and perhaps a day's work and a valuable scientific specimen will be lost. After a successful turning, trim the excess jacket and matrix away and plaster the bottom of the block the same way you did the top, after wetting both the matrix and the sides of the upper jacket. On a block that is too big to be carried to your truck by one man (large fossils are often found in places where your transportation can't go) plaster in carrying poles and find a friend.

The specimen is now ready to transport home after one vital step which you may have taken while the plaster was drying. Write up your notes on the locality with all the possible geographic and geologic data you can assemble. Without this information to accompany it, your fossil is a curiosity, not a worthwhile scientific specimen.

COLLECTING INVERTEBRATES

Invertebrates usually present different problems from those confronted when collecting vertebrates. However, a large clam or ammonite may require the methods outlined above.

In some areas invertebrates weather out of limestones and other rocks and literally cover the ground as a lag concentrate when gentle erosional forces remove the finer-grained material, leaving the fossils behind. In unconsolidated sand you may find invertebrates littering the surface, or you may want to screen for them. This can be done by building screens with half-inch mesh, putting legs on them, and then shoveling in the sand.

Often shells will be partially decalcified by the action of ground water. In this case the shell must be hardened with shellac, Glyptal, or white glue and later removed in a chunk of matrix. Then it may be prepared in the laboratory. Sometimes the shell will be completely removed, leaving a natural mold of the outside and an internal mold or steinkern (rock kernel) in the inside. You may want to pour plastic into the void and cast a replica of the shell or save the two molds in their original condition.

It is not unusual to find fossils in nodules or concretions. If you know that certain types of concretions often contain fossils while there is no surface indication of the fossil, then the geology hammer or a single jack is the only exploration tool. Gather up a bag of concretions and start pounding away.

If the specimen is partially exposed in the concretion you should wrap it up and take it back to your laboratory to be prepared. There, with chisels, Zipscribes, Mototools, or Vibrotools you can do a proper job.

In the late Cretaceous Pierre Shale and the marine facies of the Fox Hills Formation in South Dakota and adjoining states, ammonites and nautiloids may occur in nodular masses. These can be dug out of outcrops or picked up on plowed fields. Taken back to the laboratory they can be readily prepared into beautiful specimens.

In some bluffs along the Snake River in southern Idaho there are cannonball-size nodules weathering out of the silts. Most of these were formed around buried crayfish where the crayfish acted as a catalyst, causing the precipitation of the calcite which cemented the silt into a hard mass. These can only be found by cracking the nodules in two with a hammer.

COLLECTING LEAF, INSECT, AND FISH IMPRINTS

Many fossils are formed as flattened impressions or carbonaceous smears in fine-grained silts and shales. In this case you patiently sit surrounded by slabs of rock and split them along the bedding planes with a knife or a thin chisel.

The Green River fish at Fossil Butte National Monument and other localities near Kemmerer, Wyoming are both carbonaceous impressions and petrified bone. When split on a bedding plane near the fossil the vertebral column will show up as a series of bumps like a buried string of beads. Careful work with a needle will often reveal a beautiful fish.

At Florissant Fossil Beds National Monument, beautiful examples of insects and plants are typical examples of this type of preservation.

Anywhere you find fine-grained lake or volcanic ash deposits, there is a good chance that you can find beautiful leaf and insect imprints. Caution: The paleobotanists and paleontologists who study insects do not want their study specimens gooped up with preservatives.

Don't put a hardener on the specimen. Put it on the surrounding matrix if you must, but leave the specimen in its natural state.

WASHING MATRIX FOR SMALL SPECIMENS

I briefly mentioned washing and screening for specimens above. There are two approaches to this task. Some people prefer to sack up large amounts of matrix and take it to the laboratory for processing. This is the best approach when the matrix won't readily break down by washing and has to be broken down by using kerosene or some other method.

Others wash the matrix at the nearest available stream. If you make your screens with wooden frames, you can tie them into rafts, anchor them to the bank, and let the matrix soak as long as necessary. If the matrix breaks down readily, it can be washed and screened by hand without prolonged soaking.

Using the in-the-field method, the screened matrix can be laid out on a tarp to dry. When dry it can be sacked in plastic garbage bags, lined gunny sacks, or cardboard grocery boxes. Back at the laboratory it might be well to soak and rescreen the matrix to further reduce its bulk.

Figure 22. Washing and screening matrix to recover very small fossils.

Be sure to thoroughly clean your screens between localities to avoid the problem of contamination.

One area I collected years ago was a well-cemented stream channel in the rocks near Wounded Knee, South Dakota. We had seen and collected tiny bone fragments, teeth, and jaws from the surface and sides of the exposed channel. We took up 500 to 600 pounds of this matrix and hauled it back to California. There we used the kerosene method of breaking down the matrix. After washing and screening we had reduced the concentrate so that it just filled a 5-gallon kerosene can. After that came the long process of sorting out the fossils under a binocular microscope a half-teaspoon of concentrate at a time.

At another locality we did a preliminary washing in a nearby river, dried and bagged the residue, and then took it back to the laboratory where we would put batches to soak over a weekend and further reduced the concentrate. Four hundred pounds of residue from the first washing finally reduced to five 3-pound coffee cans of concentrate.

When floating rafts of screens, beware of the weather. Your river may rise due to a storm many miles away. I watched the White River in South Dakota rise 5 feet one day when we hadn't had any rain for 50 miles from camp. A friend was cleverly floating his screens in a lagoon on the west coast of Baja, California when a storm went over the bar and the breakers wiped out his operation.

A HOME FOSSIL LAB

If your ambition doesn't run to preparing large skeletons, whale, mammoth, or *Triceratops* skulls, you can assemble a very effective fossil lab on the top of a sturdy card table. The other limitation is the good nature of your spouse on the matter of dust and flying rock chips covering the living room rug.

The basic requirement, after the card table, is a wooden tray about 2 feet square and 2½ to 3 inches deep with a solid plywood bottom. Put 2 inches of sand in the tray and cover that with a loose piece of burlap cut to size. The burlap can be lifted off to dump matrix chips as they accumulate, and if you accidentally chip off a piece of bone or shell, you can find it on the cloth without having to sift through your sand. When gluing, you can fold the burlap at the far side of the tray and position the bonded pieces in the sand, which will hold them in place until they dry.

Three containers for liquids (alcohol, glue, and shellac or thinned Glyptal) can be readily made from the plastic squeeze bottles from catsup or mustard sold in variety stores and supermarkets. Run a small, round paintbrush through the spout tip so that the bristles are about a half-inch from the bottom of the bottle when the top is screwed down tight.

The alcohol is useful for softening matrix, assisting natural partings, and sometimes for distinguishing bone from matrix when the two have similar colors and the matrix is very fine. I'm old-fashioned and still believe that white shellac cut by three or four parts alcohol to one of shellac is a good utility hardener and crack sealer. For adhesives you may want two kinds. A plastic bottle of white glue with a brush through the spout for application is handy, but I'm still a Duco Cement fan. You can use the cement as it comes from the tube you buy in your local store, or buy it in gallon cans and transfer small amounts to your glue bottle as needed. Duco can be thinned with acetone, but I prefer it as it comes from the tube or can. For most gluing it can be applied with a brush, but for tiny things the tip of one of your parting needles will hold just the right amount.

Among the expendables, you should have paper towels and plaster. Plaster of Paris is expensive if you are going to use more than just a little dab. An investment in 100 pounds of casting plaster at your local lumberyard will save money in the long run and keep you supplied in both field and laboratory for some time. In the lab you can mix it in paper cups with a tongue depressor or ice cream stick, or get rubber mixing cups and thin steel spatulas through any dental supply house.

Light to work by is important. This is pretty much a matter of individual choice. The old-fashioned student's gooseneck desk lamp is very handy, as you can maneuver it around to almost any position. A sand bag on the base sometimes helps avoid a crash when you try to stretch it beyond the limits of equilibrium. Small high-intensity lamps can now be purchased at many discount houses or other stores. These are handy for small work, particularly if it has to be magnified, but they don't give the overall illumination you might like. If your affluence knows no bounds, you could get a good fluorescent desk lamp that clamps to the table and may be positioned with a cantilever or telescoping arm. I got along for years with the old gooseneck type and have only a few scars on the backs of my hands where I touched the hot bulb while concentrating on a delicate fossil.

To round out the basic equipment you should have an assortment of small sandbags. They are handy for bracing pieces being glued, for holding down your lamp when it has to be placed in a precarious position, and as bases for working on tiny pieces that require a firmer support than the burlap-covered sand in the tray. There aren't many Bull Durham smokers anymore, but those small tobacco sacks make excellent sandbags. A few scraps of leather, canvas, or denim and a short tour on a sewing machine will give you a variety of bags. I'd suggest four or five, ranging from little 2-inchers up to at least one that is about 8 inches long, 4 or 5 inches wide, and about 2 inches thick. If you need larger sizes for more ambitious projects, you might approach your kindly neighborhood banker for some canvas money bags.

If your fossils are small, you might want some sort of a magnifying device. Again this is a matter of choice. A binocular microscope is nice (and expensive), but you can generally get along with any of the various gadgets that clip onto your glasses or fit over your head. A jewler's loupe is virtually worthless, as the undistorted area is so small, and you have to hold your head so close to the work.

Don't forget to have a sharp knife or a hacksaw blade for cutting plaster jackets. A good hooknose linoleum knife is best—not one of the cheap ones from your local five and dime or discount paint store, but one of professional quality.

This leads us to one final basic tool which should always be at your elbow—a small whetstone. Sharp tools are really the one great trade secret to good, easy fossil preparation.

We have now listed most of the basics. How about the actual tools for chipping or scraping away the matrix? This again is a matter of individual preference, and you will have to evolve your own favorites from the basic stock suggested here. Brushes, hammers, chisels, and needles are all necessary. The problem is to choose or make individual tools that fit your hands and feel comfortable to the touch.

Brushes

Several assorted brushes are necessary for good preparation. A selection of the small, round artist's type is useful for dabbing alcohol, acetone, or even water on small surfaces to help natural partings or to float out fine grains of silt or sand. Soft 1-inch paint brushes can be used to gently whisk away loose matrix and should be used to overcome the impulse to blow on the specimen. If you must blow, use a baby syringe (available at any drug store) instead of the full power of your lungs. A particularly handy item is a dentist's scratch brush, a bundle of fine steel wires held together in a short piece of rubber hose about ⅜ inch in diameter. These are ideal for the final gentle cleaning of a bone or shell surface.

Hammers

Chiseling is often needed to work off or reduce thick matrix. Various types of small hammers are available with heads weighing but a few ounces. I prefer the metalsmith's chasing hammer, which has good balance and a broad head that makes missing the chisel impossible for all but the most clumsy. This type of tool is available from rock shops and catalog suppliers that handle jewelry-making tools.

Chisels

Small cold chisels are available commercially, but even the smallest of these is usually too large for fine fossil preparation. A good source of raw material for making your own chisels is your kindly family dentist. Ask him to save his broken tools for you. These, when the tips are broken off about a half-inch from the handle, can be shaped into suitable chisels on a small shop grinder without destroying their temper. The handles may be cut off with a hacksaw to appropriate length. When working a specimen, hold the chisel near its tip, with your hand firmly braced on the fossil or the sandbag. The sharp tip should be held an eighth-inch or so away from the specimen. When you strike the chisel with the hammer you will get a bouncing impact and a more delicate control than you would if you held it against the matrix and tried to control the impact solely by the weight of your hammer stroke.

Needles

While you are begging your family dentist for old tools, be sure to ask him to save his old drill burrs for you. These can be shaped into good needles on the small whetstone that should always be in your basic stock of tools. Other needles can be made from small awl blades or even, if all else fails, from sewing needles. A variety of pin vises are available at hobby shops, and you will have to find several that fit your hand and your needs. A homemade handle carved from a piece of dowling is often your most handy and useful tool. This way you can get a perfect fit for your hand and you might make one that will become your faithful companion for years.

A very handy needle can be made from a large, round dental burr. Carefully grind down two sides so the tip is flattened into a round blade with a sharp, round leading edge. This is a good tool for removing matrix from a surface, and the round edge prevents the digging and scratching that sometimes results from a pointed needle hitting a soft spot on the fossil's surface. Grip the needle handle between your first and second fingers with the thumb holding it firmly, very much like using a pencil but with the fingers extended so the point is an elongation of your fingers. All the movement of the needle should be done with the three fingers. The hand should be firmly planted on the sand bag, sandbox, or on the fossil itself. After a short time you will develop a large and interesting callus or lump on the inside of your middle finger which will last for years. It won't do you any harm, and it can become a fine conversation piece if you can find a way to inject it into a boring conversation.

These are the basics. From here you can add to your stock in any way you wish. Macdonald's Workshop Law states: Tools will increase in quantity and complexity in direct proportion to the thickness of the wallet and the imagination and persuasiveness of the buyer when explaining the need to the spouse.

A Mototool, Vibrotool, or Zipscribe are handy additions if you can convince your spouse that all will be lost without them. These handy little gadgets grind or tap away matrix in very little time, and often do a better job than hand tools. Even more expensive and exciting is an overhead motor with a flexible shaft and chuck for holding small drills, controlled by a foot pedal under the table. An ultraviolet light

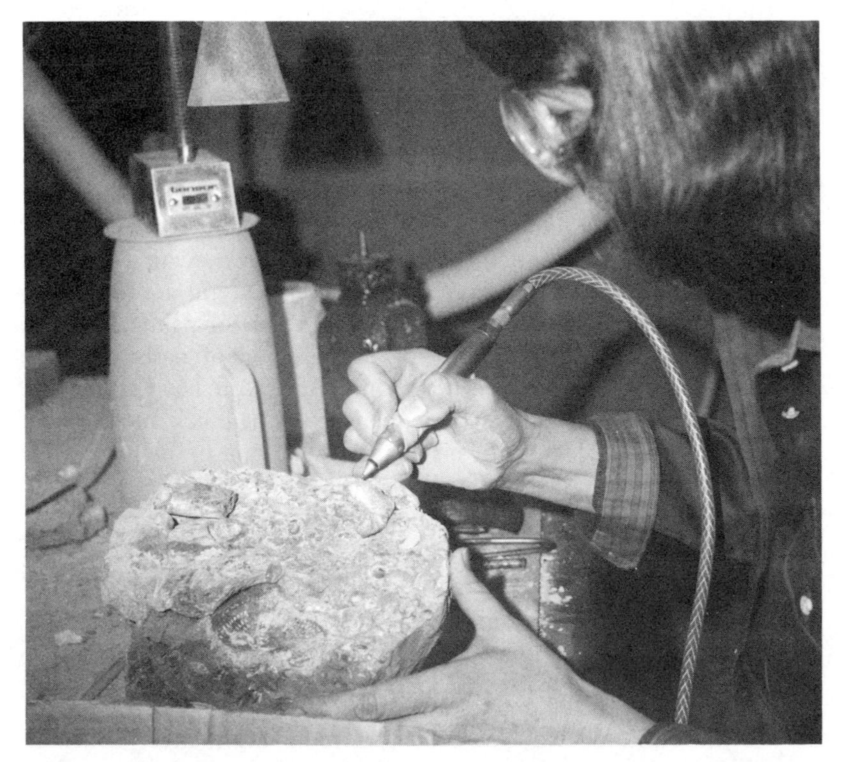

Figure 23. Preparing small ammonites with a pneumatic chisel.

will sometimes help differentiate matrix from bone. In fact, I know of several very important specimens that could never have been prepared if it weren't for ultraviolet illumination.

If you believe that the synthetic resins and fiberglass are the only molding media which match your talents, then you should have an assortment of throwaway brushes, tongue depressors, and paper cups for handling these materials. I would be the first to admit that large fossils cannot be molded, cast, or reconstructed without the use of fiberglass, but for most small workshops it is really not necessary.

Your table-top fossil lab is now ready for use. As you work on your collection you will find that new needs appear; their solution will be in your own imagination and ingenuity. You may find that you could even discard some of the things you thought were indispensible in the beginning.

ABOUT PREPARATION

You've collected a fossil and assembled your laboratory on a card table. Now there's no excuse for not going to work and doing some preparation.

Open the plaster jacket on the bottom of the fossil. It is always a good idea to start with the down side as there is always less weathering of the specimen there. When the underside has been prepared it will be a solid base for the cleaning of the exposed and possibly weathered side.

Carefully cut the bottom half of the jacket away with a hacksaw blade or your sharpened linoleum knife. Be very careful not to dig into the fossil. It often helps the preparator if the collector paints diagrams or comments on the block, showing the position of parts when he numbers it after finishing the plastering. After you have gotten the jacket off without damaging the specimen, settle the block into the sand tray so that it is stable. You may have to use one or more of your assorted sand bags to hold it firmly in place.

Before touching the block with any kind of tool, study it carefully to see if there is any bone exposed, what the cracks are doing, and if there is any crumbling. Brush the loose dirt away with a paintbrush or blow it gently away with a baby syringe. Any exposed bone should be lightly shellacked and any broken pieces glued back into place or put aside for later gluing. Be very careful to protect the broken edges, and do not gum them up with glue or thick shellac.

If the block is solid, you can begin removing matrix with small chisels. Being very careful not to dig into the bone, start near an exposed piece and slowly work away from it to expand the exposed area. If you are lucky, there will be a good parting between the bone and the matrix. If not, parting can sometimes be improved by gently wetting the interface at the exposed edge with either water or alcohol.

To be completely successful in preparation, you should know what the specimen is and what the finished product is going to look like. Only in this way can you be sure that you won't blunder into a vital area with your chisels. Of course, the fossil may be a mystery, or much distorted, or many elements may be jumbled together, so great care is always needed. Work slowly and remove only small bits of matrix at a time, so that if you do damage something, it will be minimal.

If the matrix is soft and easily crumbled, then you will not have to use chisels but can do all your preparation with awls and needles. This is often no less difficult than chiseling, as the bone may be as soft as the matrix and require constant hardening as it is exposed. It is often a good idea to have two projects going so that one can dry while you work on the other. As in the field, working on a glued or shellacked specimen before it is dry is an invitation to disaster.

As you expose the fossil, you will have to decide whether to set broken pieces aside or glue them back immediately. I favor gluing them back unless they are projections that are going to be constantly in the way and in danger of rebreaking. Your progress and what tools you use are going to be a matter of judgment as you go along. The main thing to remember is that the secret to fossil preparation is slow but continuous work.

When you have finished the bottom half of the block, the time has come to turn it over and work on the upper side. You now have the problem of exposed bone which must be the base for further work. I suggest making a plaster base for further preparation. Cover the side you have already prepared with wet paper towels. Wad up toweling to fill any depressions so the top of the wet towel surface is fairly smooth. Mix up a batch of plaster and spatula it over the wet towels to make a smooth base. When the plaster has set, very carefully turn the specimen over and bed the plaster into the sand tray just as you did when you first began to work.

The procedure on the upper surface is just the same as on the bottom. Remove the top of the plaster jacket, or as much as can be taken off without something falling apart. If the block is crumbly, remove the jacket a section at a time. The upper surface may have more cracks and be somewhat weathered. Missing pieces of bone can later be filled in with plaster. Reconstruction can be done either with plaster or papier-mâché. When the matrix has been completely cleaned away a final touch-up can be done with the dental scratch brush and the specimen lightly shellacked.

Sometimes a skull is preserved with the jaws tightly closed. A paleontologist will want to examine the crowns of the teeth, so they should be separated. This is a process that takes a great deal of patience and some skill. Clean away all the matrix from around the jaws and from the palate. Clean out the matrix in the zygomatic arches, around the ascending ramus, and around the bullae or ear bones, so that only where bone is firmly pressed against bone or teeth against teeth is there any connection between the jaws and the cranium. You may have to break away the zygoma to do this job completely.

When you have removed all the matrix possible, the time has come to try your patience and your skill. Begin slowly painting water or alcohol along the lines of bone-to-bone contact so the liquid will be absorbed by the thin film of remaining matrix. Continue this until you can see dampness appear on the other side of the joint. Then as you continue wetting carefully, very gently with a needle pry on the jaw. It may take several days of careful work to separate a well-cemented jaw from the cranium. Be sure to work on the jaw hinge at the same time you work between the teeth.

When the specimen has been completely prepared, you may wish to make a stand for it. This can be done with an appropriate wooden block for a base and a support of bent wire or copper tubing. A piece of copper or iron tubing can be split into a number of strands at one end. These strands can then be shaped to run along the basicranial area of a skull and along the palate. Side runners can hook under the bullae to give lateral support, and additional hooks can be made to support the jaws. When this plumber's nightmare is finished, the tube can be cut to an appropriate length and the solid end set in a hole in the top of the wooden block.

Another method of making stands is to bury the specimen in sand so that just the bottom is showing. Using thin boards, build a retaining wall around the fossil. Cover the bone and sand with a thin layer of wet paper towel molded to the contours of the specimen and fill the retaining wall with plaster. When the plaster sets you will have a block with an inlet support. Tool the surface of the block for prettiness and stain or paint it to suit. If you want to add a finishing touch, punch the data for the fossil on plastic labeling tape and lay this face down on the wet towel in an appropriate place. The plaster block will pick up the label and give a nice professional look to your stand.

The specimen is now ready for display after you have properly identified it. Remember that it should be recorded in a catalog and numbered in an inconspicuous place with permanent ink or paint. All of its geologic and geographic data should be entered in your catalog to keep the specimen from being just a curiosity.

If it is a vertebrate fossil, it certainly should be shown to a vertebrate paleontologist at the nearest university or museum. He will verify your identification and may very well consider it important enough to go to the locality himself. Or he may ask you to donate the specimen to the museum. In this case you should do so; it may be a valuable scientific item which will shed additional light on our knowledge of the history of life. If it is a heretofore unknown animal, the paleontologist will name it for you when he formally describes and names it in the scientific literature. In this case you will have attained a form of immortality because your name will be known as long as science survives on this earth. If you do give it to a museum, you should insist on being given a record of your gift and a good cast in exchange. This is only fair, and any good museum will honor such a request.

CASTS

Speaking of casting, you might want to make casts of your specimens to give to friends or to send off to museums for identification. There are many methods of molding and casting, but I'll just suggest three simple ones which can be done at home without a lot of equipment or potential hazard.

Simple casts of simple items can be made with modeling clay squeeze molds. Work up a suitable amount of modeling clay into a soft but not gooey mass. Flatten it out on a table with a roller. A handful of talcum powder in a handkerchief can be used for dusting the surface of the clay to keep it from sticking to the roller. If you want to mold just one face of the specimen, you can press the fossil into the well-powdered surface to make an impression and then gently pull it out without distorting the clay. Pour in some plaster and your cast is completed.

If you want to do a complete mold, press the specimen halfway into one piece of well-powdered clay, then powder the top of the specimen and press another well-powdered piece of clay on top. With your fingers, firmly work the two pieces of clay together so that they press against each other and the specimen very tightly. Then take a spatula or a knife and trim the edges of the clay blocks into square vertical walls. Now carefully pull up the upper block. Be careful not to distort it in any way. It should part with the specimen and the lower clay block if there are no undercuts and if the interfaces have been well powdered.

Next pull the specimen out of the lower clay block. Your mold is now ready to use. Mix plaster to a whipped cream consistency and fill both cavities to overflowing. Place the two clay blocks together so that the cut edges meet exactly as they did when you made the mold. Press the blocks together so the excess plaster runs out the sides. When the plaster sets, remove the cast and gently wash up the clay mold for reuse. This is not a process that is going to work perfectly the first time you try it. I'd suggest you practice with something fairly indestructible—a simple plastic toy, a chess piece, or other small object.

A similar process can be used to make a many-piece mold with plaster. By greasing the specimen and using clay dams, a complicated mass of interlocking plaster molds can be made which can be provided with a pouring hole and then held together with rubber bands. Excellent casts can be made this way, but it requires a great deal of patience and work.

Liquid latex makes very good molds and is easy to use. Again, build clay walls along proposed joints in the mold and paint the specimen with the parting agent recommended by the manufacturer of the latex you use. Paint on a layer of latex within each area and let it dry. Continue this until the mold is almost a quarter-inch thick. You can strengthen it by painting in an occasional layer of unbleached muslin.

When the first part is done, you may want to make a plaster false mold over the latex to make it hold its shape when full of plaster. Turn the specimen around and make the other pieces of the mold in the same way. With care, such molds can be used many times. Several years ago, one of my preparators, Mary Odano, made a mold of a 7-foot *Triceratops* skull in only three pieces using this method. She used fiberglass for the false mold, as plaster would have been much too heavy to handle.

Here is the final method for doing little complicated items such as small mammal dentitions. Perfection Dental Cream, one of the molding agents used by dentists, comes in boxes of small packets with a measuring cup for mixing the proper proportions with water. This material makes a very fine flexible mold and sets up rather quickly. A fine-grained hard casting plaster used by dentists can be used with these molds if you want something better than a plaster of Paris cast.

In the 1950s I was doing a study of two groups of primitive hooved mammals, the Leptochoeridae, little cat-sized piglike animals from the Oligocene, and another primitive artiodactyl family, the Anthracotheres, the size of domestic pigs and the ancestors of the hippopotamuses. I traveled to all the museums in the country that had specimens of these animals and made photographs, measurements, and casts of many of them. The Perfection Dental Cream produced quick and easy molds which I could make without a mess and then cast in my hotel or motel room at night. The museum I was working for in those days still has a complete set of casts of the important Leptochoerid specimens in the country's museums.

I would like to emphasize again that many fossils are scientific rarities and that they do not belong in private collections. A museum association can offer you research facilities; safe, permanent storage of specimens; and perhaps access to preparation tools and supplies you could never afford. Close contact with a museum and its staff can be a rewarding experience, and your key to admission is cooperative collecting and sharing with the museum and future generations.

SHIPPING SPECIMENS

The day is long gone when you could go into a one-horse town with a railway depot and ship a box of fossils home by rail. The "less than carload lots" bit is nothing but nostalgia. If you are in the position where you have to ship your collection by truck or by ship, you will have to build boxes or crates. Boxes may be padded on the inside with excelsior, hay, straw, banana leaves, or whatever comes to hand. I've used them all, and they all work.

I remember a delightful July day at Winslow, Arizona when another student and I were building crates and boxes on the depot dock. Just across the square from the depot was a house of ill repute. Whenever a potential customer walked by, the place came alive and the juke box roared out its music. Up the street a block was a saloon that served a big schooner of beer for 10¢: *This* was the Lorelei that pulled us from the dock and up the street about once every half-hour in the 110-degree-in-the-shade temperature. Who wants a whorehouse under those conditions?

I was told years ago that the box shown in Figure 24 was known as a Weyerhauser pack. The following lumber is needed to build this box:

Top and bottom:	2	1″ × 6″ × 36″
	2	1″ × 12″ × 36″
Sides:	2	1″ × 6″ × 37⅝″
	2	1″ × 12″ × 37⅝″
Ends:	2	1″ × 6″ × 15⅝″
	2	1″ × 12″ × 15⅝″
	2	1″ × 4″ × 18¾″

When this box is built and covered as shown in the illustration it will require a nail puller to take the top off. When packing the box,

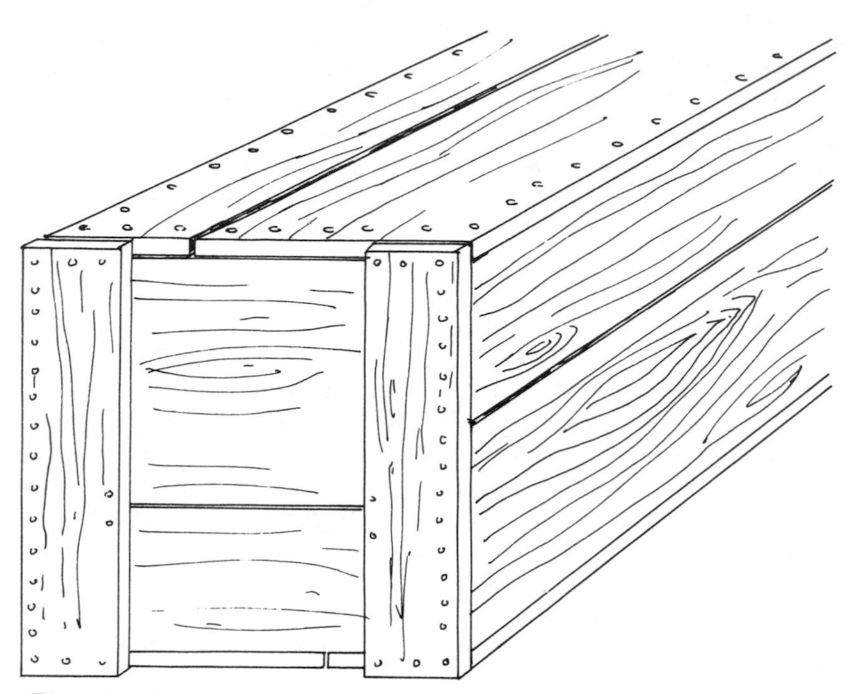

Figure 24. Construction of the Weyerhauser pack.

heavy specimens should be secured by tightly nailed partitions so they do not destroy smaller specimens.

FIELD NOTES

Here is a suggested format for field notes. I've used this style at several museums and it works well.

- Name and year to be entered at upper left-hand corner of each page.
- Page number (numeration to begin with first entry of any calender year and to continue until the last entry of that year) to be entered in center of top of page.
- Each daily entry will be started with a heading line.
- Date at left margin, place where notes are written (locality at end of day) to right of margin. Heading line to be underlined for the whole width of the page.
- When a daily entry continues on to a new page, a new heading line will be made on the first line of the new page.

- New localities will be listed and will be described both geologically and geographically. One method of numbering will consist of the author's initials; a "V" for vertebrate localities, and an "R" for rock sample localities; the last two numbers of the year; and the number of the locality for the year. JRMV635 would be Macdonald's fifth vertebrate locality discovered in 1963.

Macdonald
1963

Pg 26

15 July Wounded Knee, South Dakota

Spent A.M. prospecting Sharps fm. in outcrops near junction of Sharps Cutoff and Gooseneck Rd. Collected material at two sites

JRMV635 on west side of Sharps cutoff, 300 yds. N. of junction w/ Gooseneck Rd.-SE 1/4 of SW 1/4, Sec 17, T39N, R43W, Shannon Co. 90-100' below contact w/ Monroe Creek fm.

JRMV636 on south side of Gooseneck Rd. opposite junction w/ Sharps cutoff.-NE 1/4 of NW 1/4 and NW 1/4 of NE 1/4, Sec 20, T39N, R43W, Shannon Co. 80-90' below contact w/ Monroe Creek fm.

17
SHARP'S CUTOFF
JRMV365
GOOSENECK RD
JRMV636
20

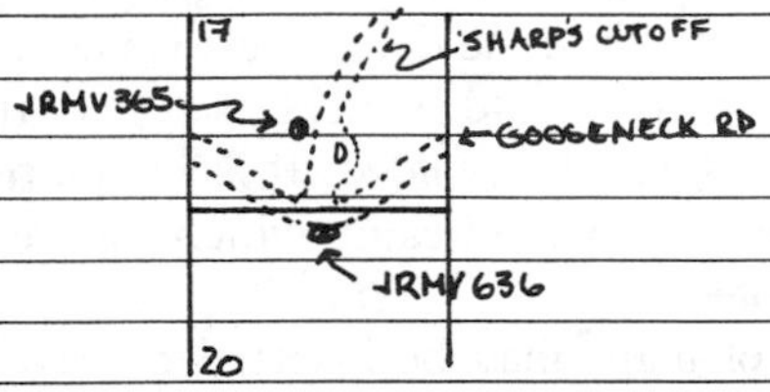

JRM-
840 JRMV635
Paleocastor SKULL
Sharps fm. Jones

Figure 25. A sample field notebook page.

- If an old locality is being recollected the assigned number is to be used. If the locality has been numbered by another museum, that museum's number will be used and a geographic and geologic description given in the notes.

  SDSM (South Dakota School of Mines and Technology, Museum of Geology) V592 – 70′ above base of Monroe Creek fm in 10″ layer of coarse sand – Center of SW of Sec. 3, T. 38 N., R. 43 W., Shannon Co., South Dakota – Porcupine Butte Quadrangle S.D. Geol. Surv.

Specimen field numbers consist of the author's initials—or party chief's initials if someone else is keeping the notes and the assigned number. Numbers should be on a lifetime basis for the individual assigning the numbers.

Specimen entries will be made on three lines:

Field number	Locality number
Name of specimen	Character of specimen
Formation collected	Name of collector
JRM 840	LACM 1892
Palaeocastor	Skull
Sharps fm.	Jones

Many specimens of a similar character may be given a single field number when coming from the same locality.

Judgment should be used in the assignment of field and locality numbers. Specimens from the same horizon in a small area should all be assigned the same locality number.

Sketch maps, geologic cross sections, hypotheses, and pertinent data should be entered in the field notes. General information about the activities of the party should be kept as part of the record. The "Dear Diary" concept should be avoided. It is a good practice to keep a sheet for a "guest book" in a field camp which can be inserted at the end of the season's notes.

Sections of maps may be inserted as pages; X-section paper or other special papers may be used as special sheets for illustrating special features.

References to standard texts, individual papers, maps by name and/or number, comments of visiting experts, and any other informative data should be entered if they will add to the pertinent data on the problem. If particular personnel or administrative problems affecting collecting in a certain area are encountered, they should be documented.

LOCALITY DESCRIPTIONS

Descriptions may follow the model format and include as much information as possible. In the event that the required information is not completely available at the time of number assignment, space shall be left for the information and a pencil note of the fact should be made.

Locality number (underlined)	*Locality name (underlined)* *Age of deposit* *Name of formation*
	Legal description with map reference (county, quadrangle, aerial photo, etc., and date of issue) and applicable stratigraphic information. Reference to field notes and field locality number of original collector.
6499	*Mission Pit* Clarendonian. Ogallala fm. Stream channel deposit lying on the Hemingfordian Rosebud formation. NE¼ of Sec. 35, T. 40 N., R. 28 W., Mellette Co., South Dakota. Geology of the Mission Quadrangle, S.D.G.S., 1963. Macdonald, 4 July 1964, locality JRM V6410

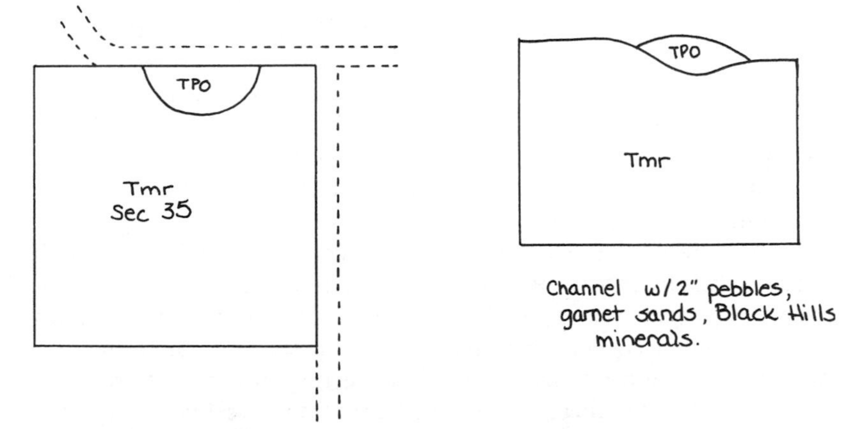

Figure 26. Geologic map and cross section to go with the locality card or catalog.

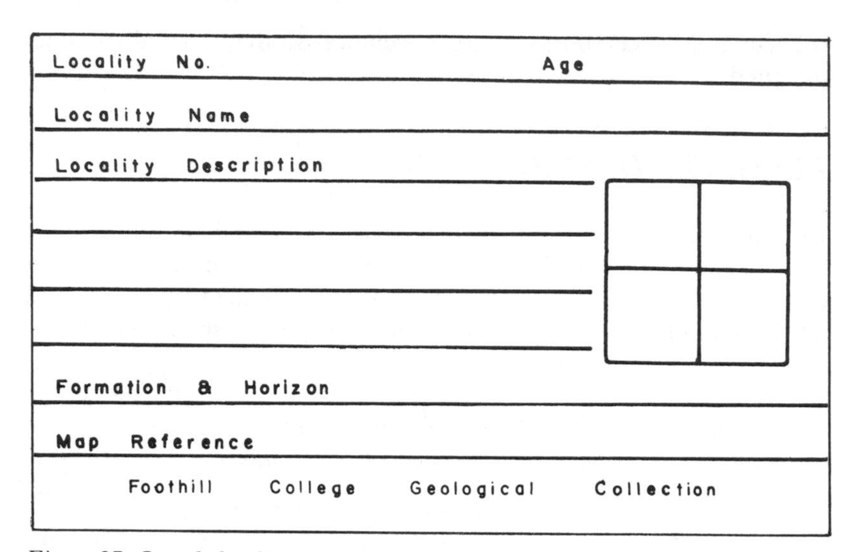

Figure 27. Sample locality card. An alternative to the method shown in Figure 28.

CURATION

No job is completed until the paperwork is done. Taking good field notes is not enough. After the specimen has been prepared and ready for display, or stored away, it should be cataloged if you want to maintain a professional standard.

Two catalogs should be used: a locality catalog to record your localities and a specimen catalog to record the items in your collection. Catalogs may be kept on cards or in notebooks. To help those that come later, a consistent system should be used. You can never record too much information. Without this information your discoveries are only curiosities. I have a curator friend who received a beautiful ammonite from the widow of a collector. The only data for the specimen was a neatly printed statement on one side, "From further upstream" [!]. The specimen would be useful in a teaching collection, but valueless in the museum's study collections.

The locality card's format that I've used in the past is a simple system, and any entry is open-ended so you may add as much additional data as you wish (See Figure 28). You may either use your field locality number for the locality or assign a new number. In a large museum where there are a number of people collecting in different areas and assigning their numbers in the field, a museum locality number is a must, but when an individual collector is keeping the notes, then the field number should suffice.

Number	Location
Name	
Locality	
Formation & Age	
Source	
Collector	
Value	
Description & Remarks	
Foothill College Geological Collection	

Figure 28. Sample catalog card. The "Location" space in the upper right-hand corner, filled in in pencil, tells where the specimen is stored.

You may wish to maintain a map file with your localities marked on the maps. These files could include topographic maps, geologic maps, air photos with the sites marked by pin pricks and the locality number written on the back side by the circled prick. Each state highway department issues county maps with scales ranging from a quarter-inch to the mile to an inch to the mile. These are very useful when nothing else is available. In Brazil I once used a gasoline company highway map to record localities.

Instant-type photographs of sites taken while specimens are being collected are also very useful.

The specimen catalog should have a format that will assign a specimen number. Some museums start numbering specimens with "1" and keep going. A different set of numbers should be used for plants, invertebrates, and vertebrates. Another method is to start a new sequence each year. Thus the first number used in 1975 would be "751." In addition to the specimen number, there should be places for the name and number of the locality, the name of the taxon the specimen represents, what the specimen is—shell, steinkern, right jaw, left maxillary, complete skeleton, etc.; the geologic age of the specimen; and the collector.

When you have put these data together, put both specimen and locality numbers on the specimen. Make a neat patch on the specimen with white model paint and when it dries put the specimen number and

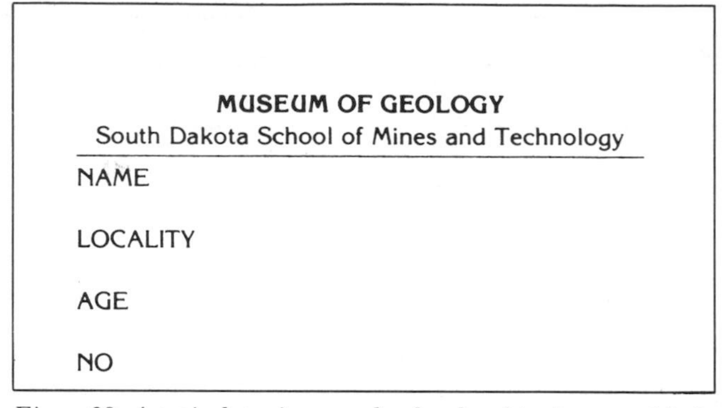
MUSEUM OF GEOLOGY
South Dakota School of Mines and Technology
NAME
LOCALITY
AGE
NO

Figure 29. A typical specimen card to be placed in the tray with the fossil.

the locality number on the patch with black india ink. A little shellac on the ink will help preserve it. Write the numbers as a fraction with the specimen number on top. You may wish to use specimen cards to be placed in the tray with the specimen or attached as a tag. These small cards, about the size of a business card, could carry the specimen and locality numbers, the taxon, the age, and the item.

If you care for your specimens in this way, your collection will have a permanent value.

LARGE FIELD PARTIES AND FIELD TRIPS

When I have conducted field trips with 20 or more students we have gone either by bus or by caravan. Buses have certain advantages, but there are many places you cannot take a bus where a car will go. One driver asked me, "Does the company know that we drive off the highway?" This was just after he had gotten stuck in a small sand dune.

When going by caravan I have used a van or a carryall from the college with a covered rental trailer as one of the vehicles. The remainder have been either school cars or private automobiles. By putting all of the camping, baggage, and field gear in the trailer the cars are roomier and can carry a capacity load.

Gasoline and oil were either included in the course fee or were purchased by the riders in each vehicle. The owner didn't chip in and was given enough course money at the end of the trip for a tank of gas, car wash, and sometimes a tuneup.

A CB radio in each car makes control and lecturing en route much easier. It is recommended that those without CBs purchase toy walkie-talkies. These are all tuned to channel 14. Put the cars with walkie-talkies just behind the lead car and there is rarely any problem in communication.

While traveling, breakfast and supper are eaten in restaurants. Lunch may be either a tailgate picnic from the general lunch supplies or eaten at fast-food places. It's been my experience that if you set up a picnic lunch in a town park, most of the group will head for the nearest McDonald's or its equivalent.

Spend your traveling nights at roadside rests where allowed. Otherwise look for a forest campground or a pullout.

If your group is traveling by bus, eat all your meals in restaurants. The driver and I always slept in the cargo bins under the bus so we didn't have to bother with tents.

The daily schedule while traveling might be like this:

6:00	Up all hands.
6:30	On the road.
First Town	Stop for breakfast for an hour then back on the road.
ca. 12:00	Stop for lunch for an hour and then back on the road.
ca. 6:00	Stop for supper for an hour and then on to your overnite.
Twilight	Stop for the night. Have a short meeting after everyone gets their gear organized. Review what was seen during the day and outline the next day's activities.

You will want to schedule morning and afternoon pit and sandbox stops. This will depend on the fuel range of the automobiles and the physiology of the group. This is one place where the CB radios come in handy.

Daily Schedule While in Camp

6:00	Reveille.
6:30–7:30	Breakfast and clean up. Get ready for the field.
7:30	Into the field.
12:00	Lunch.
1:00	Back into the field.
5:00	Return to camp.
6:30–7:30	Supper and clean up.

7:30 Show-and-tell, write notes, wrap and label packages.

10:00 No more talking or noise in camp.

During show-and-tell the leader will examine, evaluate, and discuss the specimens collected during the day. He will also write his field notes and assign field numbers to the specimens. The specimen wrapper will wrap and number the specimens as the leader gives them to him.

On my field trips grades are based on attitude, field notes, and serendipity. If someone finds a spectacular or scienceworthy specimen they get an "A" if their attitude and field notes are acceptable. I try to impress on my students that they should always be ready to "grab the bucket." If they see something that needs doing—do it!

Personal Equipment for Camping with a Group

- Tent
- Sleeping bag and pad or air mattress
- Field clothes with a warm jacket
- Boots or stout shoes
- Bathing suit—I've never gotten used to women wearing bikinis while prospecting or working in a quarry, although this does lead to some spectacular pictures.
- Towel
- Toilet articles
- Cup, knife, fork, and spoon; a dinner plate if you can't abide paper ones
- Notebook
- Camera—if desired
- Poncho or raincoat—sometimes nice to have
- Insect repellant

CAMPING AND COOKING

I've been on trips where camp cooking was done on an individual or cooking group basis. The cooking groups were usually the occupants of each car. Frankly, I don't like this method. The diets are not always the best and the sanitary situation often deteriorates. In my opinion it is better to have a single camp mess.

The general equipment for the camp and cooking for 20-25 people should include:

- 2 Three-burner camp stoves—preferably with a propane converter that will attach to a trailer-type propane bottle

- 1 Two-burner camp stove—either gasoline or propane
- 3 Folding tables—2 for cooking, 1 for serving
- 2 Grills with grease troughs and a drain hole that will fit on the three-burner stoves. These can be readily made from a sheet of quarter-inch aluminum.
- 2 Canning kettles with lids
- 2 Large (1-gallon or more) coffeepots
- 2 2-quart plastic pitchers with tops
- 1 Cutting board
- 2 Large plastic bowls for salad
- 1 Wooden chopping bowl and chopper
- 2 Metal buckets
- 2 Folding chairs
- 2–4 Hot pads
- 3+ Camping ice chests
- 2 Toilet brushes
- 5–6 5-gallon water cans
- 1 10-gallon Igloo-type water container with spigot
- An assortment of knives, spoons, forks, turners, potato peelers, etc.
- 1 Latrine screen and an entrenching shovel
- 1 Canvas tarp for covering stores when in the field and at night

DUTIES ASSIGNED TO STUDENTS

- Cargo Master: Supervises the loading and off-loading to the trailer.
- Ice Box Person: Keeps ice chests clean and drained and buys ice as needed. Leader will keep him supplied with money.
- Gunga Din: Keeps water cans filled.
- Miner: Digs garbage pits and latrine slit trenches as needed. Slit trenches should be about 4 feet long and the width of an entrenching shovel. I once had a coed come to me after the Miner had dug an 18-inch-wide slit trench in the newly erected latrine. She reported, "Somebody doesn't understand girls!" Leave the spoil and the entrenching shovel in the latrine so people's contributions can be buried—but not too deeply as it uses up the slit trench too fast.
- Stove Person: Keeps track of the fuel supply and lights the stoves in the morning and puts the coffeepots on one three-burner stove and two buckets of water on the two-burner stove. One bucket will have liquid soap poured into it. The toilet brushes should be

by this stove. Pots, pans, dishes, and utensils are washed in the soapy water and rinsed in the clear water. Be sure the water is boiling and that everything is rinsed thoroughly. In my camps the stoves are lit at 6:00 A.M. The Stove Person then pounds on my trailer or camper and then makes the rounds waking up the party. He also sets up the two-burner stove for supper.

- Breakfast Layout Person: Sets up the serving table for breakfast. Mixes powdered milk and keeps track of supplies. Makes list of needed items for the shoppers, who will usually be the Leader and the Supper Cooks. Puts away the tablegear and packaged goods after breakfast so he/she will know where it is. Don't forget to put out toilet paper on the breakfast table.
- Breakfast Cooks (one for each grill): Also make lists for shopping.
- KP Pusher (pick someone who is older and mean, preferably an ex-first sergeant from an infantry company: Assigns students to kitchen police on a rotating basis. The Cooks are exempt from this duty.
- Kitchen Police: Wash utensils, stoves, grills, and tables; dispose of trash.
- Lunch Person: packs food and utensils for lunch; lays out lunch on tailgate, flat hood, or picnic table; cleans up after lunch; keeps track of lunch supplies so the shoppers can keep up to date.
- Supper Cooks (may be the same as the Breakfast Cooks): Prepare supper and make shopping lists for the shoppers. The Leader will assign supper menus on the basis of shopping availability; for example, the night after you've been through a town is the night to have fresh meat.
- Supper Salad and Layout Person: Prepares salad and lays out the serving table for supper. Cleans up after supper and makes shopping lists for the shoppers.
- Specimen Wrapper: wraps, numbers, and packs the day's collection at show-and-tell.
- Map Keeper: Records localities on maps during the day and at show-and-tell.
- Grid Artist: (when working in a quarry or whenever else desirable)—Makes grid maps of the excavation based on 1-meter grid system.

FIELD TRIP MEALS

The following tried and true menus were designed for a class of 20 students and 2 instructors. Groups vary: Some will eat you out of house and home while others eat like canaries. The first day or two should give you an idea about quantities for any particular group.

Here are two breakfast menus. We always alternate from day to day, and there is always dry breakfast food for thems that likes the stuff. One coffeepot is for boiled ("prospector's" or "cowboy's") coffee, and the other has hot water for the brewing of tea, cocoa, and powdered coffee.

Breakfast #1

- Eggs—44
- Hash—4 1-pound cans
- Bread—2 large loaves
- Oleo—1 pound in a tub
- Jam—3-pound tin
- Coffee—large jar of instant
- Sugar—in dispenser
- Coffee lightener—large jar
- Milk—1 gallon of dry nonfat
- Cocoa—large box
- Dry cereal—individual boxes, assorted
- Oranges—22

Breakfast #2

- Hotcakes—4 pounds of complete mix
- Bacon—2 pounds
- Syrup—1 pint (serve in a dispenser)
- The remainder of the meal (coffee, etc.) is the same as Breakfast #1.

Lunch is a make-your-own-sandwich affair. Here is a list of things we lay out. Tell everyone to bring their cups into the field, as there is always punch of some sort; we recommend Wyler's over the other brands.

- Bread—at least 2 large loaves, 30 slices each
- Jam—buy in 3-pint tins
- Peanut butter—buy in 3-pound jars
- Spam or equivalent—2 cans
- Vienna sausage—5 cans
- Sardines—4 to 5 cans
- Kipper snacks—3 to 4 cans
- Wylers's punch—2 gallons
- Crackers—1-pound box
- Cookies—2-pound bag
- Apples or other fruit—22
- Pickle relish, catsup, and mustard

In general, supper is a one-pot meal. Below are some simple field-tested menus. These are easily modified, and you can probably come up with any number of your own. Usually on the last night out it's steaks and beer (pop for the non-beer drinkers). One year I had a Japanese student as Main Supper Cook. There was suddenly a lot of class to these mundane dishes. I've always tried to allocate a couple of hundred dollars for my Lead Cook on a two- to three-week field trip. It is richly deserved.

Dinner #1 Beef Skillet Mexicana

- Hamburger—8 pounds
- Onions—3
- Pitted olives—1 6-ounce can
- Green pepper—1

Chop up these four things and start browning in a skillet or two. When brown and hot, add:

- Corn—2 1-pound cans, drained
- Tomatoes—3 1-pound, 12-ounce cans, partly drained
- Chili powder—4 to 6 tablespoons or to taste
- Salt and pepper

Heat these with the meat and onions until the liquid boils. Add:

- Minute Rice—A 1-pound, 12-ounce box (makes 24 servings of ⅔ cup each). Leave on a low fire for 2 more minutes after recommended cooking time, then take off and let stand in a warm place for 5 minutes and serve.
- Salad—3 heads cabbage
- Dressing—1 pint oil and vinegar with seasoning mix
- Fruit—5 large cans
- Bread—2 large loaves
- Milk—1 gallon
- Coffee—1 large jar
- Cookies—1 2-pound bag
- Pream, sugar, salt, pepper, etc. from supply

Dinner #2 Spaghetti

- Spaghetti—24 to 30 ounces
- Hamburger—7 to 8 pounds
- Sauce:

  Tomato paste—4 12-ounce cans
  Seasoning mix—7 packages
  Water (see instructions on box)
  ¼ cup oil

- Salad
  - Cabbage—3 heads
  - Tomatoes—3 1-pound, 12-ounce cans
  - Dressing—1 pint with seasoning mix
- Fruit—5 cans, 29- or 30-oz. size
- Bread—2 30-slice loaves
- Milk—1 gallon
- Coffee—1 large jar
- Cookies—2-pound bag
- Pream, sugar, cocoa, jam, salt, pepper, etc.

Dinner #3 Stew

- Stew—7 cans, 1½ pounds each; add ketchup and a chopped onion or two
- Salad:
  - Cabbage—3 heads
  - Tomatoes—3 cans, 1 pound, 12 ounces each
  - Dressing—1 pint with seasoning mix
- Fruit—5 cans, 29- or 30-ounce size
- Bread—2 30-slice loaves
- Milk—1 gallon
- Coffee—1 large jar
- Cookies—a 2-pound bag
- Pream, sugar, cocoa, jam, salt, pepper, etc.

Dinner #4 Tuna Stew

- Noodles—24 ounce package; cook this until tender and add:
  - Tuna—7 12½-ounce cans
  - Cream of asparagus or celery soup, not thinned—2 to 3 cans
  - Peas or green beans—4 1-pound cans

  Break up the tuna in the noodles and add the soup and peas or beans. Cook until hot through.
- Salad:
  - Cabbage—3 heads
  - Tomatoes—3 1-pound, 12-ounce cans
  - Dressing—1 pint with seasoning mix
- Fruit—5 29- or 30-ounce cans
- Bread—2 30-slice loaves
- Milk—1 gallon
- Coffee—1 large jar
- Cookies—2-pound bag
- Pream, sugar, cocoa, jam, salt, pepper, etc.

Dinner #5 Hot Dogs on Buns

- Franks—7 packages of 10; or 7 pounds
- Buns—40+
- Relishes—mustard, pickle relish, catsup from lunch supply
- Salad:
    - Cabbage—3 heads
    - Tomatoes—3 1-pound, 12-ounce cans
    - Dressing—1 pint with seasoning mix
- Fruit—5 cans, 29- or 30-ounce size
- Bread—2 30-slice loaves
- Milk—1 gallon
- Coffee—1 large jar
- Cookies—2-pound bag
- Pream, sugar, cocoa, jam, salt, pepper, etc.

Dinner #6 Hamburgers with Mashed or Hash Brown Potatoes

- Hamburger—10 pounds
- Instant hash browns or mashed potatoes—1-pound package
- Soup sauce—7 cans cream of mushroom soup, or something
- Salad:
    - Cabbage—3 heads
    - Tomatoes—3 1-pound, 12-ounce cans
    - Dressing—1 pint with seasoning mix
- Fruit—5 cans, 29- or 30-ounce size
- Bread—2 30-slice loaves
- Milk—1 gallon
- Coffee—1 large jar
- Cookies—2-pound bag
- Pream, sugar, cocoa, jam, salt, pepper, etc.

Dinner #7 Hamburger with Chili and Noodles

- Hamburger—7 pounds
- Chili with beans—6 1½-pound cans
- Noodles—2 12-ounce packages or 1 24-ounce package
- Salad:
    - Cabbage—3 heads
    - Tomatoes—3 1-pound, 12-ounce cans
    - Dressing—1 pint with seasoning mix
- Fruit—5 big cans, 29- or 30-ounce size
- Bread—2 30-slice loaves
- Milk—1 gallon

- Coffee—1 large jar
- Cookies—2-pound bag
- Pream, sugar, cocoa, jam, salt, pepper, etc.

Dinner #8 Franks in Pork and Beans

- Franks—7 packages of 10; or 7 pounds cut up in pork and beans
- Pork and beans—2 3½-pound cans or slightly more
- Salad:
  - Cabbage—3 heads
  - Tomatoes—3 1-pound, 12-ounce cans
  - Dressing—1 pint with seasoning mix
- Fruit—5 cans, 29- or 30-ounce size
- Bread—2 30-slice loaves
- Milk—1 gallon
- Coffee—1 large jar
- Cookies—2-pound bag
- Pream, sugar, cocoa, jam, salt, pepper, etc.

SHOPPING LIST FOR FOOD AND COOKING SUPPLIES

Buy house brands or no-brand (generic) items when possible. The quality is good and the price is better.

I always try to buy all of my nonperishable items at home base as they are generally less expensive than out in the field where you often depend on small-town grocery stores for supplies. I plan all of my menus before I leave and purchase accordingly. The fixings for each supper are prepacked in individual boxes so all that has to be added are the perishables. Other boxes are allocated for condiments, lunch materials, paper goods, and breakfast and supper serving table items. Pack groceries in cardboard boxes and save the boxes for packing fossils.

Condiments and such

- Salt
- Pepper
- Sugar
- Catsup—in "family"-size jugs
- Mustard—in plastic squeeze bottles
- Chili powder
- Spaghetti sauce seasoning mix
- Salad dressing seasoning mix

- Oleo—in plastic tubs
- Vinegar—in gallon jugs, plastic if possible
- Cooking oil—in gallon plastic jugs
- Jam—in 3-pint cans
- Peanut butter—in quart jars
- Tomato paste
- Pickle relish—get a big jar

Beverage

- Coffee
- Instant coffee
- Coffee lightener
- Tea bags
- Cocoa
- Wyler's punch in 45-ounce boxes
- Powdered nonfat milk

Fresh Meat and such

- Bacon—thin-sliced
- Hamburger—regular or low-fat, but don't keep for more than a day
- Frankfurters—at least two per person. Serve as quickly as possible.
- Eggs—large in plastic cartons if possible. The paper cartons melt in the ice boxes, so if you can't get plastic cartons wrap your egg cartons in freezer bags.

Canned Meat

- Corn beef hash—buy by the case
- "Spam"—a generic name for several brands
- Corn beef
- Sardines
- Vienna sausage—I lived on this for 10 days in New Guinea once, but there is no accounting for taste.
- Kipper snacks
- Tuna—12½-ounce cans or bigger

Dried Foods

- Minute Rice—buy the 1-pound, 12-ounce boxes
- Spaghetti—30-ounce packages

- Noodles—24-ounce packages
- Instant hash brown or mashed potatoes—1-pound packages

Fresh Produce

- Onions—5-pound bag
- Cabbage or lettuce—cabbage keeps better and is cheaper. Out in the boondocks lettuce may be fist-sized, expensive, and not always very good.
- Potatoes—if you want to use "live" potatoes instead of dehydrated
- Green peppers

Bakery Goods and Such

- Bread—mix white, whole wheat, and rye until you know what the group eats. Pack in boxes or else you will be eating bread balls.
- Crackers—in some places in the West you can buy these in cans. They are great for packing washing concentrates and small fossils.
- Cookies—an assortment in 2-pound bags
- Dry cereal—in individual boxes
- Pancake mix—get your favorite brand in 3- to 4-pound bags or boxes
- Hotdog buns—you know your group better than I do.

Canned Goods

- Fruit—buy #10 cans if possible. Figure one per day.
- Beef stew—1½-pound cans
- Mushroom soup
- Cream of celery soup
- Cream of mushroom soup
- Peas—1-pound cans
- Baked beans—#10 cans
- Chili with beans—#10 cans or 1½-pound cans.
- Pork and beans—#10 cans or 1½-pound cans
- Pitted olives
- Corn—not creamed—1-pound cans
- Tomatoes—1-pound, 12-ounce cans by the case

Paper goods and soaps

- Plates—buy 200 or 300 to start with.
- Napkins—get the multihundred big-size package.
- Paper towels—get them from the college if you can; otherwise buy a box.

- Toilet paper—it's good for both fossils and bottoms, so buy a bunch.
- Plastic freezer bags—for leftover food and fossils
- Liquid dish soap

Fuel

- Propane
- Stove gas

BASIC CAMP RULES

Keep your "home" neat—cover your things in the morning, as it may rain while you are out of camp.

Be ready to move out as soon as the breakfast residue is cleaned up.

No littering. If you find trash, pick it up—don't leave it for someone else.

Beer drinking permitted in the evening but not during field or traveling hours. (Try to enforce this one, Fearless Leader.) I was with a class on a beach south of San Francisco one day when a bus from an unnamed San Francisco college pulled into the parking lot. The instructor swung off with a half-gallon jug filled with a red liquid. He led his charges down to the beach and each of them had their own half-jug. I wonder at the quality of the pedagogy that day.

Leave gates as you find them. Close them when closed and leave them open when they are open.

Don't play Don Dumbardo with the livestock. Leave the critters alone.

No radio or tape deck playing within hearing distance of the camp. Play your own instrument as long as it doesn't annoy anyone.

Quiet after ten o'clock. One year one of the women students annoyed her neighbors because she giggled for quite a while every night after "lights out."

These trips are experiences in human relations. It must be a cooperative effort.

Remember that *Boobus americanus* is rampant throughout the land. Don't join that species.

Also remember, when you see something that should be done, grab the bucket and do it!

Chapter Five

THE ORGANIZATION OF LIVING THINGS: THE LINNEAN WAY

*I*n addition to being a junk collector, man is a compulsive classifier. Everything must be categorized and pigeonholed where it can be retrieved when wanted. Probably the earliest categories were "good" and "bad." "Good": you can eat it. "Bad": it can eat you. Later classifications became increasingly sophisticated; eventually the whole field of classification was absorbed into the science of taxonomy.

As modern science began develop in the eighteenth century, the most outstanding classifier of living things was the Swedish medical doctor Carl von Linné, known formally as Carolus Linnaeus (1707–1778), who became the most prominent botanist of his time and the founder of the system of biological classification which is in worldwide use today. Of course, Linnaeus did not invent his system out of whole cloth without having earlier systems to build on, but he did develop a methodology that could be applied to all living organisms. Linnaeus believed that all living things had been divinely created and could be only slightly modified by later events. For this reason his classification does not readily indicate the evolutionary relationships of organisms through time. His system divided living things into groups, each group being based on the presence of a single characteristic shared by every member. Each organism received a double name, a generic and a trivial or specific name, and for this we call it a *binomial system*. Linnaeus' "empire" included all living things; indeed, being alive was the single characteristic possessed by all its members. Once named, organisms were fitted into positions within a hierarchial or ranked arrangement which included the following levels:

Empire (the most inclusive grouping)
Kingdom
Class
Order
Genus
Species
Variety (the least inclusive grouping, which is no longer used in the Linnean sense)

Today we have added a number of categories and no longer use the level of empire:

Kingdom
Phylum ("Division" is commonly used by botanists)
Class
Cohort
Order
Family
Tribe
Genus
Species

Many of these levels are further subdivided by adding the prefixes *super-*, *sub-*, and *infra-*.

In pre-Darwinian times organisms were classified archetypically—that is, by the similarity of each to some hypothetical type or standard. Since evolution has become recognized as a guiding principle of biological science and the phylogeny or family tree of organisms is better understood, classification is now based on evolutionary relationships. The science of taxonomy can be divided into studies of phylogeny (evolutionary relationships), classification (placing of organisms in categories), and nomenclature (the mechanics of naming categories).

Of all the levels of classification used in the Linnean system only the species is a real entity. The higher categories are manmade devices for indicating presumed relationships between species. This is the reason the classifications vary from author to author; each sees the higher relationships from a slightly different viewpoint. This doesn't invalidate the system, it simply indicates that there is flexibility and room for differences of opinion. Understanding progresses because of differences of opinion.

To show how the Linnean system may be applied to things besides plants and animals, we could classify the family cars:

	Car 1	*Car 2*
Kingdom	Machinery	Machinery
Phylum	Transportation	Transportation
Class	Wheeled	Wheeled
Order	Automobile	Automobile
Family	Chrysler	Nissan
Genus	Dodge	Datsun
Species	360	610

Doing the same thing with man, the family cat, and an alligator would look something like this:

	Man	*Cat*	*Alligator*
Kingdom	Animalia	Animalia	Animalia
Phylum	Chordata	Chordata	Chordata
Class	Mammalia	Mammalia	Reptilia
Order	Primates	Carnivora	Crocodilia
Family	Hominidae	Felidae	Crocodylidae
Genus	*Homo*	*Felis*	*Alligator*
Species	*sapiens*	*cattus*	*mississippianus*

(Note: Genus and species are always written in italics, or underlined; the generic name is capitalized while the specific name is not.)

From the above we can see that all three organisms are both animals and chordates—that is, animals with a stiffening structure down the back. But the alligator is a reptile, not a mammal, so here its classification diverges. The cat and the man remain together until they diverge at the ordinal level, one going to the order Primates, the other to the Carnivora. If we were to put the neighbor's dog into the system, he would stay with the cat through the ordinal level and branch off at the familial level into the family Canidae, genus *Canis*, species *C. familiaris*. (The generic name is often abbreviated after the first time it is used on a printed page.)

KINGDOM PROTISTA
PRECAMBRIAN TO HOLOCENE

This kingdom is a little like a wastebasket; the various phyla vary so much in their characteristics that they might have been thrown in simply to get rid of them. Most are plantlike, some animallike, and some are plainly undecided. All are nucleated, single-celled organisms. Those that build shells or tests are often common parts of the fossil record; some are important rock formers; many are excellent guide fossils and indicators of past environments.

PHYLUM PHAEOPHYTA
PRECAMBRIAN TO HOLOCENE

The brown algae are familiar to us as the giant kelps which may reach a length of 50 meters or more. The green color of the chlorophyll is masked by a golden-brown pigment. With holdfasts to secure them to the bottom and floats to keep them upright, these organisms flourish in shallow marine waters. The body is specialized for different functions, divided into holdfast or rhizoids, a stalk or stipe, and blades which resemble leaves. Yet none of these parts is homologous to the roots, stems, or leaves of land plants. Fossils may be impressions or carbonaceous films.

PHYLUM CHRYSOPHYTA
JURASSIC TO HOLOCENE

In this phylum are the diatoms and coccoliths, important plantlike protists and rock formers. Literally thousands of cubic kilometers of diatomite and chalk are composed almost entirely of their siliceous or limy tests.

The chyrsophytes are paleontologically important because they have solid tests (shells), they are prolific rock builders, and the diatoms are one of the ultimate sources of petroleum.

Class Bacillariophyceae
Late Cretaceous to Holocene

The test of diatoms is made of two siliceous halves that fit together like the top and bottom of a pillbox. The shells are very small, even microscopic, and are ornamented with fine lines, ridges, and pores so tiny that they are barely visible even with a good microscope. Generally the

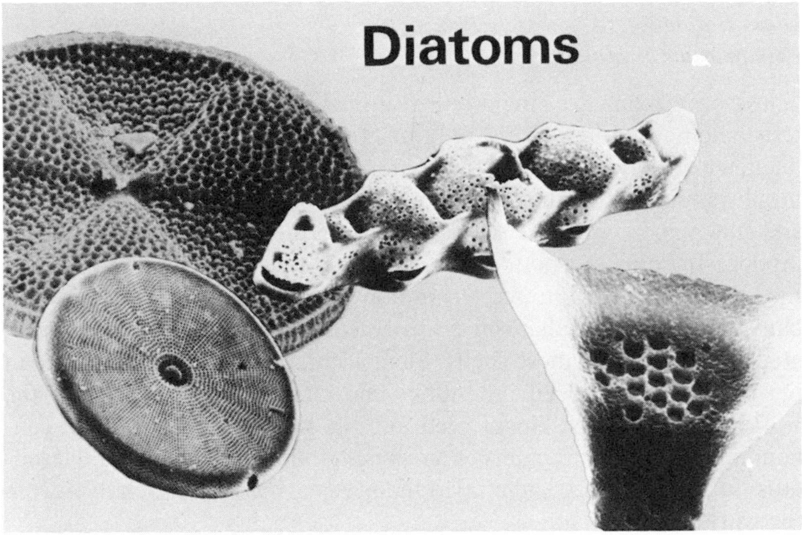

Diatoms

Figure 30. Marine diatoms from Miocene marine sediments near Lompoc, California. Courtesy of the Johns–Mansville Corporation.

marine species are round, with ornamentation radiating out from the center, while the freshwater species are canoe-shaped and bilaterally symmetrical. Diatoms are important members of the lowest part of the basic food web in the world's oceans. Vast numbers of their corpses constantly rain down to the ocean floor, where the nearly indestructible siliceous tests either form a rock known as diatomite or become one of the silica sources for a rock called chert. Both freshwater and marine diatomites are extensively quarried for use in filters, abrasives, and insulation. Of particular importance to all of us is the fact that diatoms store their food in the form of oil, which petroleum geologists think must be one of the major sources of petroleum.

PHYLUM SARCODINA
CAMBRIAN TO HOLOCENE

The familiar ameba is the prototype of these common, animallike protists. While amebas do not have hard shells for preservation, there are members of the phylum which secrete tests of chitin (a phosphatic material similar to fingernails), lime, or silica, and even some which select bits of debris from the sea bottom and build their shells by agglutination. Two groups, the Foraminifera and the Radiolaria, are important in the fossil record.

Class Foraminifera
Precambrian to Holocene

These protozoans are single-celled animallike forms which have hard tests ranging from less than 1 millimeter to more than 100 millimeters in diameter. Today they range throughout the oceans from polar to equatorial waters. Although some are pelagic (floating), most are benthonic and slowly move over the sea bottom using their pseudopodia for locomotion. In some sea bottom areas foraminifera may represent 70% of the enclosed fauna. One very common genus, *Globigerina,* is so abundant in modern seas that some 35 percent of all ocean bottoms is covered with an ooze of these shells. Such accumulations of foraminifera in the past have produced limestones and chalks in many parts of the world. The limestone blocks used to build the pyramids of Egypt contain the coin-shaped remains of *Nummulites:* the Greek historian Herodotus, fifth century B.C., identified them as petrified lentils left over from the workmen's lunches.

Foraminifera developed the ability to construct calcareous tests in the Mississippian Period, and from then on those with limy shells dominated the scene, becoming important rock builders as their tests accumulated on the sea bottoms. Three groups of foraminifera have been particularly notable rock builders at different times. In the Pennsylvanian and Permian Periods a group of wheat-grain-shaped forms known

Figure 31. Wheat-grain-shaped foraminifera (Fusulinids) which are common in marine Pennsylvanian and Permian beds throughout the world.

as fusulinids dominated. Later, in the Eocene, the nummulites were common, and today various species of *Globigerina* are forming limestones at an amazing rate, up to 1 centimeter in 260 years.

Class Radiolaria
Precambrian to Holocene

The Radiolaria were named for the radiating pseudopodia that spread out from the central test in all directions. The tests are made of silica or strontium sulfate, with some chitin often involved in the inner capsule. The tests are spherical or bell-shaped with an inner capsule containing the nucleus or nuclei and an outer jellylike body which contains the outer shell and from which the pseudopodia extend.

For sheer beauty there are few organisms to compare with the radiolarians. Their varied latticework globes and bells are artistic triumphs which can only be appreciated under a high-power microscope. Their minute size, delicacy, and great variation have made them a neglected group in paleontological studies, however. Some 900 genera and thousands of species have been described, but there is still much to be done.

THE KINGDOM PLANTAE— THE GREEN WORLD

What is a plant? A green, leafy thing? Yes, most plants are green, but there are plenty that aren't. Many organisms that are green are no longer included in the kingdom Plantae. For our purposes, it is useful to define plants as multicellular living things which, through the medium of chlorophyll, manufacture their own food from sunlight, soil nutrients, carbon dioxide, and water. The "food" they manufacture is carbohydrate—sugars and starches, in many forms and of varying complexity. They make it for their own use, but animals, including man, use it, too. Plants are near the bottom of the food chain or web; without them there would be no animal life. They need us, we need them.

DIVISION LYCOPSIDA: CLUBMOSSES, QUILLWORTS, AND SCALE TREES DEVONIAN TO HOLOCENE

The living lycopods give very little indication of the remarkable ancestry of this once-important group. Small, even inconspicuous, seldom more than a foot high, they inhabit mainly the warm and damp regions of the world. They have small, narrow leaves covering their stems, which usually grow running along the ground.

Some of their fossil relatives, however, were of tree size and so abundant that their remains in some localities make up nearly all of many great coal deposits around the world. Those trees, the scale trees or Lepidodendrales, were often more than 30 meters tall. They lived at the edges of widespread swamps, particularly during the Pennsylvanian Period, and when they died their trunks, branches, and leaves fell into the water. Instead of decaying, the vegetation in coal swamps remained intact and was slowly compressed by the weight of more vegetation and, later, overlying sediments. The volatile compounds were driven off and the mass turned to coal. After millions of years what might once have been a 30-centimeter-thick pile of leaves becomes 2½ centimeters of coal.

The scale trees (Lepidodendrales) were the forest giants of their time. Their spreading branches were completely covered with thousands of narrow, pointed leaves a few centimeters long. The leaves remained on the tree a year or so and then were shed. Where they parted from the trunk or branches a distinctive diamond-shaped scar remained, so that a denuded trunk looks like the scale-covered side of a very large fish—hence the name.

One common genus was *Lepidodendron*. Thirty meters tall or more, its branches spread out regularly to form a leafy crown. Leaves were arranged along the trunk and branches in a spiral pattern which became especially visible after the leaves had fallen and only the scars remained. Another common genus, *Sigillaria*, had leaves arranged in straight vertical rows up the unbranched trunk, which presumably ended in a fluff of long, grasslike leaves at the top. Roots of scale trees, when found alone, are assigned to the form genus *Stigmaria*. Botanists realize they may not all be related, but as they are similar in structure and cannot be assigned to a known trunk, this is a convenient way of grouping them.

DIVISION SPHENOPSIDA: HORSETAILS AND THEIR RELATIVES DEVONIAN TO HOLOCENE

Like the lycopods, members of this division were once fantastically numerous and are now reduced to a position of little importance in the plant world and are represented by one genus, *Equisetum*. At the time the lycopods were contributing their bulky remains to Pennsylvanian coal swamps, the equisetoids were doing the same—with smaller bodies, perhaps, but with greater numbers. Members of one order even attained considerable size.

Fossil equisetoids took various forms; some were little different from living species. One genus, *Calamites,* lived from Devonian to early Permian times and might have been 30 meters tall (but more often only 7 or 10). It closely resembled *Equisetum* in other details.

Another fossil group, now extinct, included the abundant genera *Sphenophyllum* and *Annularia.* Both had jointed stems and whorls of leaves like other equisetoids, but here the leaves were perhaps 2 centimeters long, sometimes wide and heart-shaped, sometimes narrow and pointed. The whole plant, with an openly branched structure, stood perhaps 25 centimeters tall.

DIVISION PTEROPSIDA: FERNS
DEVONIAN TO HOLOCENE

The sporophyte plant produces thousands of spores in sporangia which typically line the underside of the leaf. Although most fern leaves are compound (having each leaf split up into many smaller leaflets), some are entire or whole. Some bear spores on each mature leaf, some have special fertile leaves for the job. Both living and fossil forms are known to have attained heights of perhaps 18 meters, but most are of more modest height.

DIVISION PINOPHYTA: THE GYMNOSPERMS
DEVONIAN TO HOLOCENE

Two important subdivisions make up the Pinophyta. These are the Cycadicae and the Pinicae. The Cycadicae include two large groups of plants, the cycads and the seed ferns, and to one of them goes the honor of having produced the first seed. The seed ferns, then, while they had fernlike foliage, were not ferns. They were gymnosperms (the name means "naked seed"). Arising some time in the late Devonian and surviving into the Jurassic, seed ferns were so numerous during the Carboniferous that that period was named the Age of Ferns, though as it turns out, incorrectly. Important seed fern genera were *Neuropteris, Alethopteris,* and *Pecopteris* and the Permian *Glossopteris. Glossopteris* has long been cited as evidence of continental drift; it is spread throughout the southern hemisphere in a way a land plant could not easily attain unless the continents were once much closer together.

The subdivision Pinicae includes one order, the Cordaitales, which is known only from fossils. *Cordaites* appear to have been the earliest of the conifers. As reconstructed, this Carboniferous genus was a tree perhaps as much as 35 meters tall with long, narrow leaves and cones borne on special fertile branches. Descended from the Cordaitales are the modern conifers or Pinales, including such trees and shrubs as the pine, fir, spruce, and redwood. A third group includes the yews, and a fourth the ginkgos or maidenhair trees. Unimportant today, the ginkgos have a fossil record dating back into the Devonian Period. The genus survives only under cultivation today; no wild populations of *Ginkgo* are known.

DIVISION MAGNOLIOPHYTA (ANGIOSPERMAE) *EARLY CRETACEOUS TO HOLOCENE*

Angiosperms appear in the fossil record with great suddenness early in the Cretaceous, and they rapidly become the dominant element of worldwide floras.

Angiosperms range in size from the smallest and most delicate herbs to gigantic tropical trees. Their habitats vary from tropical to polar extremes, from salt marshes to alpine snowfields.

Class Lilippsida (Monocotyledonae) *Early Cretaceous to Holocene*

All members of this class produce a single leaf on first sprouting. Included are the grasses such as our domestic grains and corn, palms, bamboo, and the bulbs such as tulips, lilies, and daffodils. Most monocots have parallel leaf veins, the flower components are in groups of three, and the arborescent forms do not develop seasonal secondary wood.

Class Magnoliopsida *Early Cretaceous to Holocene*

These angiosperms produce two leaves from the bursting seed, and with the conifers they make up the bulk of our temperate forests. Most flowers and shrubs of the garden belong to this group, as do the familiar oaks, maples, elms, the domestic fruit trees, and the primitive magnolia, from which the group gets its name. The leaf veins are usually branching, and the flower components are in groups of four or five.

THE KINGDOM ANIMALIA— THE TOP OF THE FOOD CHAIN

PHYLUM PORIFERA PRECAMBRIAN TO HOLOCENE

Complete sponges are not common in the fossil record due to the ease with which the spicules scatter when the animal dies and the soft parts begin to disintegrate. Exceptions include two Silurian genera, *Astraoespongia,* a 2½- to 8-centimeter saucer-shaped calcareous sponge, and *Astylospongia,* a globe less than 2½ centimeters across. The siliceous sponge *Hydnoceras* from the Devonian is particularly common in some deposits.

Receptaculites from the Middle Ordovician is another common "sponge." This fossil is placed in the Porifera for lack of a better pigeonhole, as there is no certainty as to whether or not it really belongs here. It has been called a plant by some workers and others have placed it in a separate phylum, the Receptaculitida.

PHYLUM COELENTERATA PRECAMBRIAN TO HOLOCENE

The coelenterates are divided into three classes. The first two, the Hydrozoa and Scyphozoa, have left a meager fossil record as they do not have hard parts.

Figure 32. The glass sponge Hydnoceras *from the Devonian of western New York.*

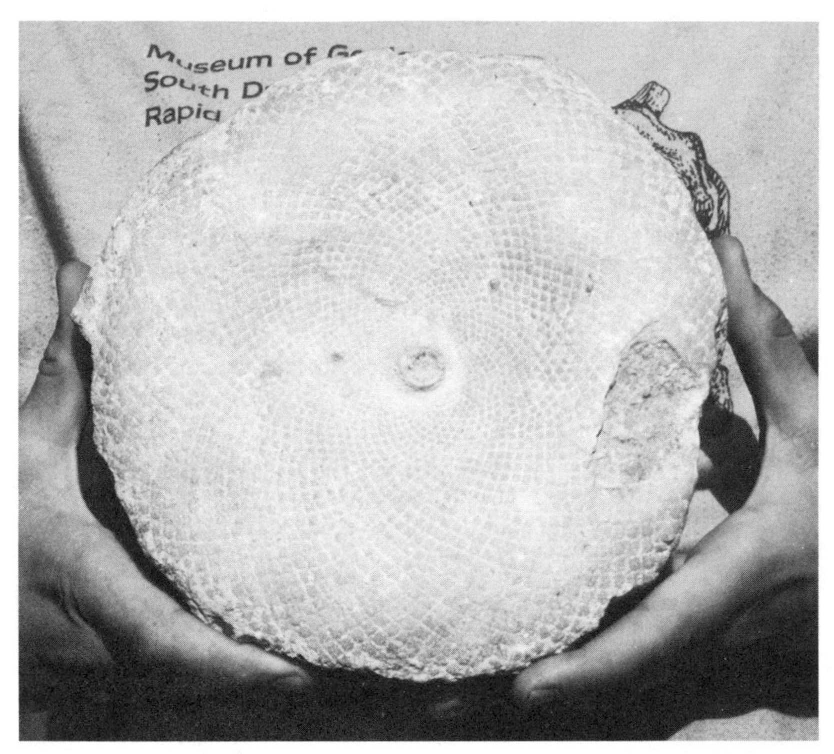

Figure 33. Receptaculites, *an Ordovician spongelike animal from the Midwest.*

Class Anthozoa
Precambrian to Holocene

These are the corals, which build external skeletons, and the sea anemones, which are completely soft-bodied. The class is exclusively marine. It flourishes primarily in shallow, clear, sunlit tropical waters.

The Anthozoa have a radial or bilateral symmetry. The body cavity is divided into pie-shaped compartments by a series of radiating mesenteries which develop in multiples of six or eight. This provides a great deal of digestive surface within a relatively small cavity. The body itself is fastened to the bottom either by a solid attachment or by holdfasts at the base of a stalk.

Corals form their shells by first laying down a horizontal baseplate on a solid surface. Then the outer shell is built up along the sides of the individual animal. As the shell grows upward, most corals build a series of radiating walls (septa) on the inside of the shell. These septa grow between the mesenteries and cause a folding of the digestive tissue which lines the body cavity. This greatly increases the digestive surface available to the animal without increasing its overall size. As the shell is built higher, the coral animal moves up in the shell and abandons the lower part. This abandoned part may be completely walled off by solid floors called tabulae or by partial floors called dissepiments.

Subclass Tabulata
Middle Ordovician to Permian

These are colonial corals which generally do not have septa. As the individual polyps grow, the lower part of the corallite is floored off by thin horizontal plates or tabulae. The more common and distinct genera include:

- *Favosites* Ordovician to Permian
- *Halysites* Late Ordovician to Silurian
- *Syringopora* Silurian to Pennsylvanian
- *Aulopora* Silurian to Pennsylvanian

Subclass Tetracorallia (Rugosa)
Middle Ordovician to Permian

The tetracorals lived in both the solitary (simple) and the colonial (compound) mode. The simple forms are referred to as horn corals; the

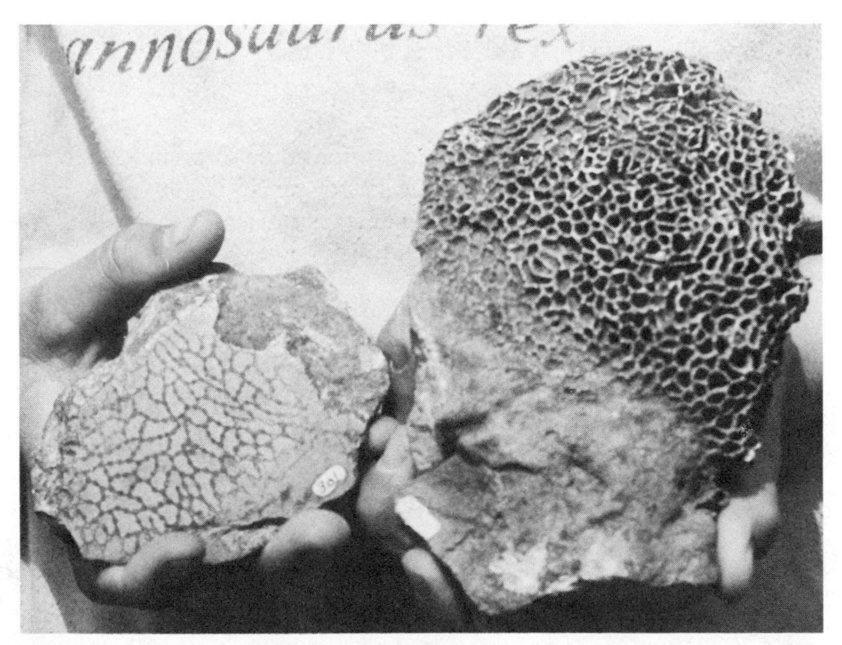

Figure 34. Halysites, *a common Silurian chain coral. On the left is a cross section and on the right an eroded boulder.*

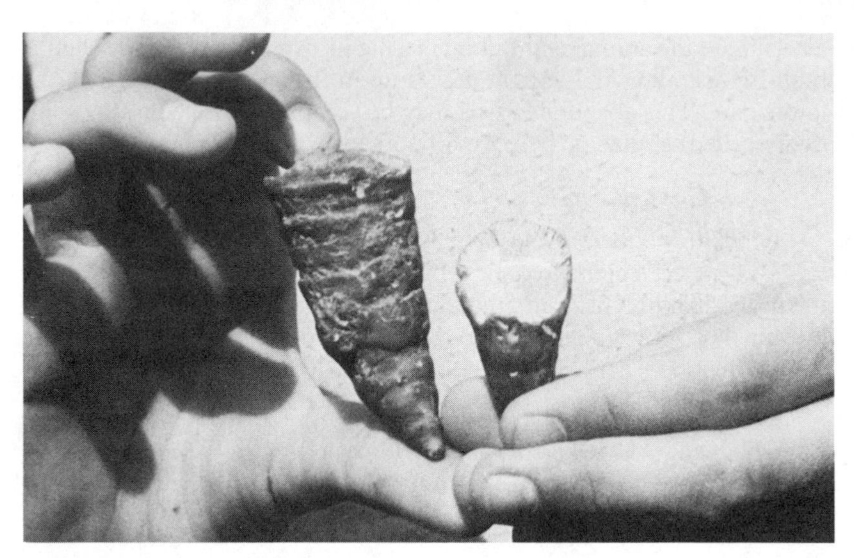

Figure 35. Rugose horn corals, very common in the Devonian.

colonial types do not have any such special designation. The subclass names are derived from the development of the septa which initially divide the calyx into quadrants and form the wrinkled (rugose) outer surface of the horn corals.

Common genera include:

(Solitary without a columella)

- *Streptelasma* Middle Ordovician to Middle Silurian
- *Heliophyllum* Early to Middle Devonian
- *Caninia* Devonian to Permian
- *Zaphrentis* Silurian to Mississippian

(Solitary with a columella)

- *Lophophyllidium* Pennsylvanian to Permian

(Compound)

- *Synaptophyllum* Silurian to Devonian
- *Hexagonaria* Devonian

Subclass Hexacorallia (Scleratinida)
Middle Triassic to Holocene

With the extinction of the Rugosa by the end of the Permian, there was a pause in the coral record until the hexacorals began to flourish following their appearance in the Middle Triassic. This is the dominant living

Figure 36. Hexagonaria, *a common colonial Devonian rugose coral.*

group, which includes the skeletonless sea anemones. The mesenteries and septa are in multiples of six. After the first six septa form between the mesenteries, others may grow between them. Thus the progression goes 6, 12, 24, and so on.

Typical hexacorals include:

- *Micrabicia* Cretaceous
- *Favia* Holocene

PHYLUM BRYOZOA ORDOVICIAN TO HOLOCENE

Literally meaning "moss animals," the name for this phylum was coined because colonies of living species encrust rocks, shells, seaweeds, and other solid surfaces like a coating of moss. Others, both living and extinct, developed colonies that resemble bushes and trees, fans, and even screws.

Typical fossil genera include:

- *Fistulipora* Ordovician to Holocene
- *Homotrypella* Ordovician to Permian, ?Triassic
- *Hallopora* Ordovician to Devonian
- *Membranopora* ?Middle Jurassic, Cretaceous to Holocene
- *Archimedes* Mississippian to Permian
- *Polypora* Ordovician to Permian

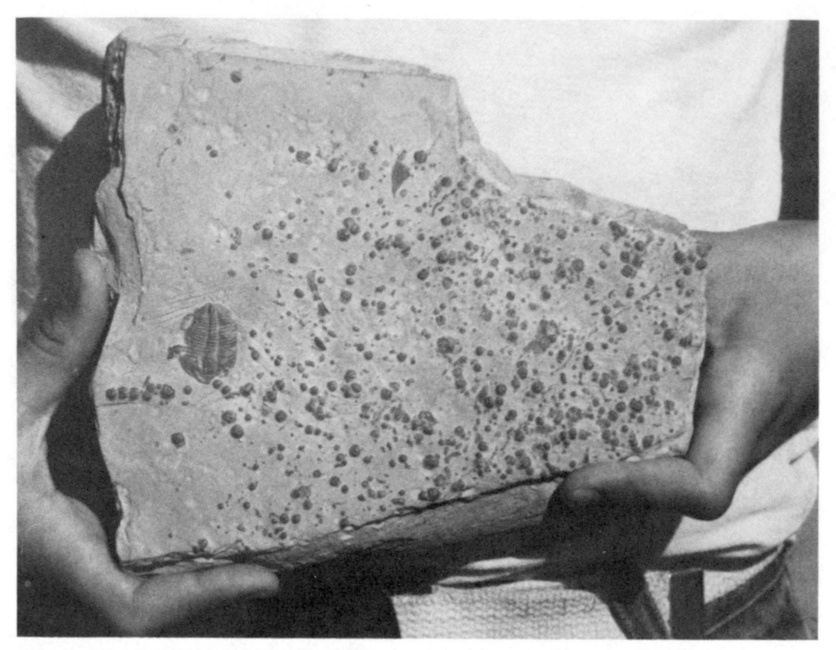

Figure 37. A trilobite and fingernail-size brachiopods from the Cambrian.

PHYLUM BRACHIPODA
CAMBRIAN TO HOLOCENE

The body of the brachipod is encased in two shells or valves which are bilaterally symmetrical with the plane of symmetry passing through both valves from the front to back. The upper, dorsal, or brachial valve and the lower, ventral, or pedicle valve differ in shape, size, and surface configuration. The shells may be made of phosphatic material or layers of calcareous deposits.

Class Inarticulata
Cambrian to Holocene

The phosphatic or calcareous valves form round low cones or teardrop-shaped cones without a hinge.

Order Atremata
Early Cambrian to Holocene

In this primitive order the pedicle opening is through both valves in a pair of notches or through a slot in the ventral valve. The living genus *Lingula* belongs to this order and is one of the longest-ranging animals with a known span from possibly the Ordovician, but certainly the Silurian, to the present day.

Other typical genera include:

- *Obulus* ?Early Cambrian, Middle Cambrian to Early Ordovician, ?Middle Ordovician
- *Trimeralla* Middle Silurian

Order Neotremata
Early Cambrian to Holocene

Here the pedicle is restricted to the lower valve. It passed through a notch that may be partially closed by a cover called the homoeodeltidium. Some become cemented to the bottom and lose their pedicle.

Typical genera include:

- *Schizambon* Late Cambrian to Middle Ordovician
- *Orbiculoidea* Ordovician to Permian
- *Schizocrania* Ordovician to Early Devonian
- *Acrothele* Middle Cambrian

Class Articulata
Cambrian to Holocene

These are brachiopods with calcareous shells and a posterior hinge that has interlocking teeth and sockets. The shells are opened by diductor muscles and closed by adductor muscles. This class may be divided into as many as eight orders. These divisions are based on the shape of the shell, the hinge structure, and the outer ornamentation.

Order Orthida
Early Cambrian to Late Permian

These shells are biconvex with the hinge line parallel to the hinge axis.

Typical are:

- *Orthis* Early Ordovician, ?Middle Ordovician
- *Platystrophia* Late Ordovician
- *Hebertella* Middle to Late Ordovician
- *Rhipidomella* Early Silurian to Late Permian

Order Strophomenida
Early Ordovician to Early Jurassic

Brachiopods with a wide variety of shell shapes with one shell usually concave and the other convex. The shell may be wider than long and the hinge line is often the widest part of the shell.

Typical genera include:

- *Plectella* Early Ordovician
- *Strophomena* Middle to Late Ordovician
- *Rafinesquina* Middle to Late Ordovician
- *Derbyia* Mississippian to Permian
- *Productus* Mississippian to Pennsylvanian
- *Juresania* Early Pennsylvanian to Early Permian
- *Dictyoclostus* Mississippian

Order Tetrabratulida
Early Devonian to Holocene

These brachiopods have a short hinge, the beaks are strongly developed, and the shells are generally smooth.

Typical genera include:

- *Terabratula* Miocene to Pliocene
- *Rensselaeria* Early Devonian
- *Pygites* Early Cretaceous
- *Terabratulina* Late Jurassic to Holocene
- *Kingena* Cretaceous
- *Stringocephalus* Middle Devonian

Order Rhynchonellida
Middle Ordovician to Holocene

Impunctate brachiopods have sharp beaks, narrow hinges, and costa or plications. Some of these show an extrordinary degree of interlocking along the anterior edge of the valves. The Devonian genera *Uncinulus, Plethorhyncha,* and *Ladogia* illustrate various ways that this is accomplished to allow a maximum opening area to the outside with a minimum of exposure to predators.

Typical genera include:

- *Lepidocyclus* Late Ordovician
- *Uncinulus* Devonian
- *Plethorhyncha* Early Devonian
- *Ladogia* Middle to Late Devonian
- *Rhynchonellina* Late Triassic to Early Jurassic
- *Rhynchonella* Late Jurassic

Order Spiriferida
Middle Ordovician to Jurassic

These brachiopods, generally with biconvex shells, may grow greatly elongated hinge lines, giving them later "horns."

Typical genera include:

- *Composita* Late Devonian to Permian
- *Athyris* Early Devonian to Triassic
- *Euryspirifer* Early to Middle Devonian
- *Mucrospirifer* Middle Devonian
- *Spirifer* Mississippian to Pennsylvanian

PHYLUM MOLLUSCA CAMBRIAN TO HOLOCENE

This large and diverse phylum includes thousands of living and extinct species which are spread throughout the ocean waters, in brackish and fresh water, and a variety of land habitats. They include the snails and their close relatives the clams, octopuses, squids, cuttlefish, the chambered nautilus, the extinct ammonites, and the lesser-known chitons and scaphopods.

Class Monoplacophora Early Cambrian to Holocene

Fossil remains of these animals with simple caplike shells were often grouped with the gastropods until living forms were found in marine muds in deep water off the west coast of Mexico in 1952. The inside of the shell has paired muscle scars running from front to back and the body is bilaterally symmetrical, a characteristic which is lost in the gastropods. It is possible that this group could be the ancestors of the polyplacoporas; perhaps the solid shell divided into segments between the muscle attachments to give the animal greater flexibility.

Class Polyplacophora Late Cambrian to Holocene

The chitons are common dwellers of the tidal zones; some reach 32 centimeters in length. They firmly attach themselves to rocks with their muscular foot. The eight overlapping armor plates that cover the back can be plainly seen beneath the leathery hide.

The fossil record is based mainly on scattered plates, but some completely articulated specimens have been found with all eight valves in position. Evolutionary improvement since the Cambrian seems to have been concerned with the overlapping of the plates and hence a strengthening of the skeleton.

Chiton is the common Holocene genus.

Class Scaphopoda
?Ordovician, Devonian to Holocene

These animals are completely surrounded by the mantle which secretes what is usually a slightly curved tapering shell open at both ends. The foot and head extend from the larger end into the mud while the tip of the small end sticks out into the water. The siphons extend out of the exposed end to circulate water to remove waste products. The shells are generally small, but may range up to 10 centimeters in length.

Dentalium (Middle Triassic to Holocene) is a typical scaphopod.

Class Gastropoda
Cambrian to Holocene

The common snails and slugs of the garden are familiar to most people and usually serve as one's introduction to the gastropods. This experience may unfortunately lead to a dislike and revulsion for the group. On the other hand, the child who has held a conch shell to his ear and "listened to the sea" has had a much better early relationship with this class of animals.

Gastropods ("stomach foot") are so called because of the muscular foot used for crawling. The head is highly organized, usually with two pairs of tentacles, a pair of eyes, and a mouth which may be at the end of a proboscis or flexible snout. The land snails have their eyes at the ends of the larger pair of tentacles.

The shell may be a simple cap or it may be completely coiled and elaborately ornamented. Some marine gastropods have a trap door (operculum) with which they can cover the shell opening when they retreat into it.

Typical gastropods include:

- *Maclurites* Ordovician
- *Turritella* Triassic to Holocene
- *Crepidula* Cretaceous to Holocene
- *Busycon* Cretaceous to Holocene
- *Olivella* Eocene to Holocene
- *Helix* Eocene to Holocene
- *Haliotis* ?Cretaceous, Miocene to Holocene

(The abalone steak is the muscular foot of this snail.)

Class Bivalvia (Pelecypoda)
Cambrian to Holocene

This class has long been known as the Pelecypoda ("hatchet foot"), but in recent years for reasons of publication priority, which may or may not be valid, the name Bivalvia has been resurrected and is coming into common usage.

Although often modified by adaptation, the pelecypods are fundamentally bilaterally symmetrical bivalves with the plane of symmetry running between the two calcareous valves. The valves are held open when the animal is relaxed by an elastic ligament or resilium on the hinge line. To close the shells the animal must use its one or two adductor muscles (which leaves scars on the inside of the shell at their attachment points). The delicious scallops, a favorite seafood, are the adductor muscles of *Pecten*. The hinge line has interlocking teeth and sockets whose arrangement is useful in identification.

Most pelecypods live in shallow water; some cement themselves to the bottom (oysters); others are attached by threadlike strands called the bysuss; some are rock borers; others wood borers; some burrow in the mud; some move about on the bottom and some even swim (pectens). Many are commercially important for food, pearls, the mother-of-pearl inner lining of the shell, or as a source of lime for industry and agriculture. The teredos (shipworms) are destructive pests because of the damage they do to wooden ships and pilings. Sometimes when eaten during the wrong seasons they are poisonous, and the giant *Tridacna* of the South Pacific has been known (one authenticated case to my knowledge) to trap divers who get a hand or foot stuck between the valves.

Figure 38. The common pecten or scallop. Found in the marine Miocene sediments on both coasts of the United States.

The standard classification of the Bivalvia divides them into 7 subclasses and 14 orders. A simpler classification suitable for our purposes may be reduced to three orders.

Order Taxodontida
Ordovician to Holocene

The anterior and posterior muscle scars are subequal, the hinge line is elongated on either side of the ligament, and the numerous teeth are taxodont or generally at right angles to the hinge. Representative genera include:

- *Nuculopsis* Silurian to Holocene
- *Yolidia* Pennsylvanian to Holocene
- *Arca* Jurassic to Holocene
- *Glycymeris* Cretaceous to Holocene

Order Anisomiariida
Ordovician to Holocene

In this order the posterior adductor muscle is large while the anterior muscle is either reduced or missing. The teeth are greatly reduced in number and may lie parallel to the hinge line. With the reduction or loss of the anterior adductor muscle the shell tends to become shortened and in the pectens is essentially bilaterally symmetrical—the plane of symmetry cuts through both shells, as in the brachiopods. In fact, pectens can be thought of as having upper and lower valves rather than right or left valves. Some pectens are able to swim through a combination of shell-clapping and siphon action. Typical anisomiariids are the mytiloids *Myalina* (Devonian to Permian) and *Mytilus,* the mussel (Triassic to Holocene); the pectenoids *Pecten* (Mississippian to Holocene) and *Lyropecten* (Oligocene to Holocene); and the osteroids *Ostrea,* the common oyster (Triassic to Holocene), *Gryphaea* (Jurassic–Eocene), and *Exogyra* (Jurassic to Cretaceous).

Order Heterodontida
Silurian to Holocene

The teeth are both parallel and at approximately right angles to the hinge line with a few large cardinal teeth at the beak and small lateral teeth toward the front and rear.

This order is divided into a number of groups with a diversity of specializations. Most of the clam shells picked up on beaches, stream banks, and lake shores belong to the Heterodontida. The marine genera *Venus* (Cretaceous to Holocene) and *Chione* (Eocene to Holocene) and the freshwater genus *Unio* (Triassic to Holocene) are among the most common.

Coralliochama from the late Cretaceous of the West Coast is an interesting representative of an unusual adaptation. The right valve grows into an elongated cone which is buried in the muddy bottom. The left valve forms a trapdoorlike covering which is exposed above the substrate. This group of bivalves, known as the rudistids, appears in the Jurassic with the left valve forming the buried cone; by the end of the period some had "turned over" so that it was the right valve which was buried. During the Cretaceous both types went through a great expansion, but by the end of the period all had become extinct.

Mya (Oligocene to Holocene), *Panope* (Jurassic to Holocene), and *Solen* (Eocene to Holocene) are soft mud burrowers who extend the siphons up to the surface of the sea bottom.

Finally, the heart clams or cockles are the Cardiids who have heavy shells with strong interlocking teeth that hold the shells together after the soft parts have disappeared. *Cardium* (Triassic to Holocene) and *Pseudocardium* (Oligocene to Holocene) are common fossil forms.

Class Cephalopoda
Cambrian to Holocene

One of the most important fossil animal groups are the Cephalopoda ("head foot"). Some 3,000 genera and more than 10,000 fossil species have been described; all known forms are marine.

Subclass Nautiloidea
Late Cambrian to Holocene

The chambered or pearly nautilus of the southwestern Pacific Ocean is the only surviving genus of this subclass. Its five species are all that is left of this group whose origin may be found nearly at the beginning of the Paleozoic. The body is housed in the living chamber of the shell. As the body grows and more shell is built, the body moves forward in the shell and the abandoned part of the living chamber is walled off with a partition of aragonite. The pattern made by the partition where it joins the outer shell wall is important in the classification of both the nautiloids and the totally extinct ammonoids. This pattern, which is called the suture, can be seen when the shells are filled with mud or some other filling and the outer shell is broken away. In the nautiloids the partitions are concave forward and join the chamber walls in smooth curves. Communication is maintained with the abandoned chambers through a fleshy tube called the siphuncle.

The earliest and simplest nautiloids had straight or slightly curved

Figure 39. A Cretaceous chamber nautilus has been cut in half to show the chamber walls and chambers. Some of the chambers have been filled with mud while others have been lined with calcite crystals.

shells. During the Ordovician there was rapid evolution, the straight forms reaching their maximum length of nearly 5 meters with a shell 30 centimeters in diameter. These are often referred to the genus *Orthoceras,* but the name is not valid as it was earlier used as a bivalve generic name. *Michelinoceras* is the proper name now for many of the species that were once included in *Orthoceras.*

After reaching this peak during the middle Ordovician, the nautiloids steadily declined in both numbers and variety. Curved-shelled types lived through the Devonian and straight-shelled forms continued through the Triassic, but since that time all known nautiloids have been coiled.

Typical nautiloids include:

- *Michelinoceras* Ordovician to Mississippian
- *Ryticeras* Devonian
- *Nautilus* Oligocene to Holocene

Subclass Ammonoidea
Late Silurian to Late Cretaceous

The ammonites are distinguished by their complicated and pleated suture patterns resulting from a folding and fluting of the septa near their edges and (usually) by having a siphuncle near the ventral side of the shell. They may have been derived from coiled nautiloids with a ventral siphuncle or from uncoiled nautiloids with a ventral siphuncle. In either case they are already coiled with a ventral siphuncle when first recognized. Some ammonities later become partially or completely uncoiled, developed turretlike coiling, doubled back on themselves, or even built shells that look like randomly tied knots. Paleontologists are still arguing about life style and life position or orientation in some of these strangely coiled forms. Many must have been bottom dwellers, and some were perhaps even attached to the bottom.

Suture types are divided into three basic groups: goniatitic, ceratitic, and ammonitic. There is an increasing complexity of folding in this series. A division of the ammonites into three orders can be based on suture types, but there is an overlapping of patterns so that the orders, while named for the main suture type, may contain species with other patterns.

Order Goniatitida
Devonian to Permian

The most primitive ammonites have goniatitic or nonserrated sutures. During the Mississippian one line develops ceratitic sutures and during the Permian ammonitic sutures are found in another group of goniatites.

Typical goniatites include:

- *Goniatites* Mississippian
- *Cravenoceras* Mississippian
- *Munsteroceras* Mississippian

Order Ceratitida
Late Devonian to Triassic

Standing in an evolutionary position between the goniatites and the ammonites, this order grades into both groups. Ordinarily the sutures have serrated lobes. Some forms are coiled only in the youthful stage and then become straight as they continue to grow.

Typical ceratites include:

- *Meekoceras* Triassic
- *Nevadites* Triassic

Order Ammonitina
Early Triassic to Cretaceous

With very complex suturing, well-preserved specimens of this order are prized collectors' items. Generally the species are completely coiled, but some are straight, turret-shaped, hook-shaped, or loosely and partially coiled.

Various coiling types include:

Coiled	*Placenticeras*	Late Cretaceous
	Sphenodiscus	Cretaceous
	Metoiceras	Cretaceous
Partially Coiled	*Scaphites*	Late Cretaceous
	Acanthoscaphites	Late Cretaceous
Straight	*Baculites*	Late Cretaceous
Hook-Shaped	*Hamites*	Early Cretaceous
Spiral	*Turrilites*	Cretaceous

Goniatitic ammonites are characteristic from the Devonian through the Pennsylvanian, when species with ceratitic and ammonitic sutures appear. During the Permian all three types are common, although the ceratites become more prominent. During the Triassic the ceratites are the major groups and goniatites disappear. Ammonitic forms do not become important until the end of the period. As the Triassic progressed, smooth-shelled forms gave way to ornamented types and some uncoiled and turreted shells appeared. By the close of the period the ceratites were gone.

The Jurassic ammonites were entirely ammonitic. They expanded in types and numbers until they reached a climax at the time of the Jurassic–Cretaceous transition. The Cretaceous was a time of diversity

Figure 40. Baculites, *a common straight ammonite from the Cretaceous.*

in coiling and uncoiling. Toward the end of the period some forms simplified the suture patterns back into ceratitic and goniatitic patterns.

The cause of ammonite extinction at the end of the era is still a mystery. After a long and complex history, with several severe reductions in numbers, the complete disappearance of the ammonites remains to be explained.

Subclass Coleoidea
Cambrian to Holocene

Except for the five living species of *Nautilus,* all of the extant cephalopods belong to this suborder. Its members have two gills, 10 tentacles (8 short ones and 2 long ones), and an internal shell if there is one at all.

The Coleoidea include the extinct *Belemitida* (Late Mississippian to Eocene), the *Sepiida* or cuttlefish (Cambrian to Holocene), the *Teuthida* or squids (Jurassic to Holocene), and the *Octopodida* or octopuses (Late Cretaceous to Holocene). Of these only the belemnites are important in the fossil record because of their massive internal axial shell. The octopuses have no shells; the cuttlefish are moderately rare in the record, although the internal cuttlebone may be preserved; and the squids have only a small, internal, coiled and chambered shell that is rarely preserved.

Though the squids commonly seen in fish markets are small, the largest living invertebrates are the giant squids of the North Atlantic Ocean, which may reach an overall length of 18 meters.

Order Belemnitida
Late Mississippian to Eocene

The belemnite internal shell has a massive, elongated, bullet-shaped guard of calcite crystals. Inserted into the blunt end is a small, chambered shell resembling that of a straight nautiloid, with a sheathing called the conotheca. The conotheca may expand forward to form a proostracum. Generally only the guard is preserved. In some Jurassic and Cretaceous marine deposits belemnite fossils accumulate on eroded surfaces as lag gravels.

Typical genera include:

- *Pachyteuthis* Jurassic
- *Belemnitella* Cretaceous

The belemnites are thought to have come from the straight nautiloids, giving rise in turn to the teuthoids and sepoids. The origin of the octopuses is unknown, but it has been suggested that, as they appeared while the ammonites were disappearing, they might be ammonites that have lost their shells.

PHYLUM ARTHROPODA
?PRECAMBRIAN, CAMBRIAN TO HOLOCENE

If diversity of adaptation and habitat utilization are the keys to evolutionary success, then the Arthropoda are the most successful invertebrates. If sheer number of living species is the criterion, then the arthropods are the most successful living animals; nearly three-quarters of the million living animal species are insects. The jointed-foot animals or arthropods are generally elongated, segmented, bilaterally symmetrical animals with a highly developed ventral nervous system, each segment with one or more pairs of jointed legs, and a body covering, either an exoskeleton of chiton or a calcareous shell. The phylum includes marine, freshwater, terrestrial, and flying forms. In size they vary from insects less than a quarter of a millimeter in length to crabs with appendage spans of nearly 3½ meters.

Classification is complex and is by necessity simplified here. In order to contain the numerous groups that may be placed in the Arthropoda without creating new phyla, the awkward term *supersubphylum* has been used as the next division below phylum.

SUPERSUBPHYLUM EUARTHROPODA
CAMBRIAN TO HOLOCENE

Arthopods with a jointed exoskeleton and the body divided into a head, thorax, and abdomen.

SUBPHYLUM TRILOBITOMORPHA
CAMBRIAN TO PERMIAN

These aquatic arthropods include a number of trilobitoids known primarily from the Middle Cambrian Burgess Pass Shale and the familiar trilobites so prominent during most of the Paleozoic. The characteristic common to all is the possession of divided legs which have a gill branch at the base of the walking or swimming legs.

Class Trilobita
Cambrian to Permian

A combination of features distinguishes the trilobites from other groups. First is the division of the body into three longitudinal lobes extending from the head to the tail—the central (axial) lobe and two lateral or pleural lobes. The second feature is the division of the body transversely into distinct head or cephalon made of inflexibly fused body segments, a flexible thorax of unfused segments which range in number from 2 to nearly 50, and a tail or pygidium of fused segments. Third is a double pair of legs on each segment except the anteriormost; a gill-bearing branch sprouts from the base of each walking or crawling leg.

Trilobites may be grouped according to various characteristics, so a generalized classification is used here. Those that appear at the beginning of the Cambrian certainly have histories that extend well back into the later Precambrian. The first appearance of the olenellid trilobites is generally used to mark the beginning of the Cambrian Period.

Order Protoparia (Olenellida) (Mesonacida)
Early Cambrian

Here the cephalon is large with crescent-shaped eyes, the facial suture is marginal, there are numerous thoracic segments, and the pygidium is reduced to the point where it is almost gone or developed into a long spine resembling the telson on a horseshoe crab.

Typical genera include *Paedeumias, Olenellus, Nevadia,* and *Holmia.*

Order Opisthoparia
Cambrian to Permian

This was the most successful group of trilobites with their long geologic range and many species during the Cambrian and Ordovician. The facial suture of these trilobites ends posteriorly at the back edge of the cephalon between the glabella and the genal angle. The eyes are located on the free cheek. They range in size up to the giant 67.5-centimeter *Terataspis* from the Devonian of New York.

Typical genera include:

- *Olenoides* Middle Cambrian
- *Paradoxides* Middle Cambrian
- *Isotelus* Ordovician

Order Proparia
Middle Cambrian to Late Devonian

Here the facial sutures extend forward from the outside of the genal angle to exit ahead of the sides of the glabella or join in front of the

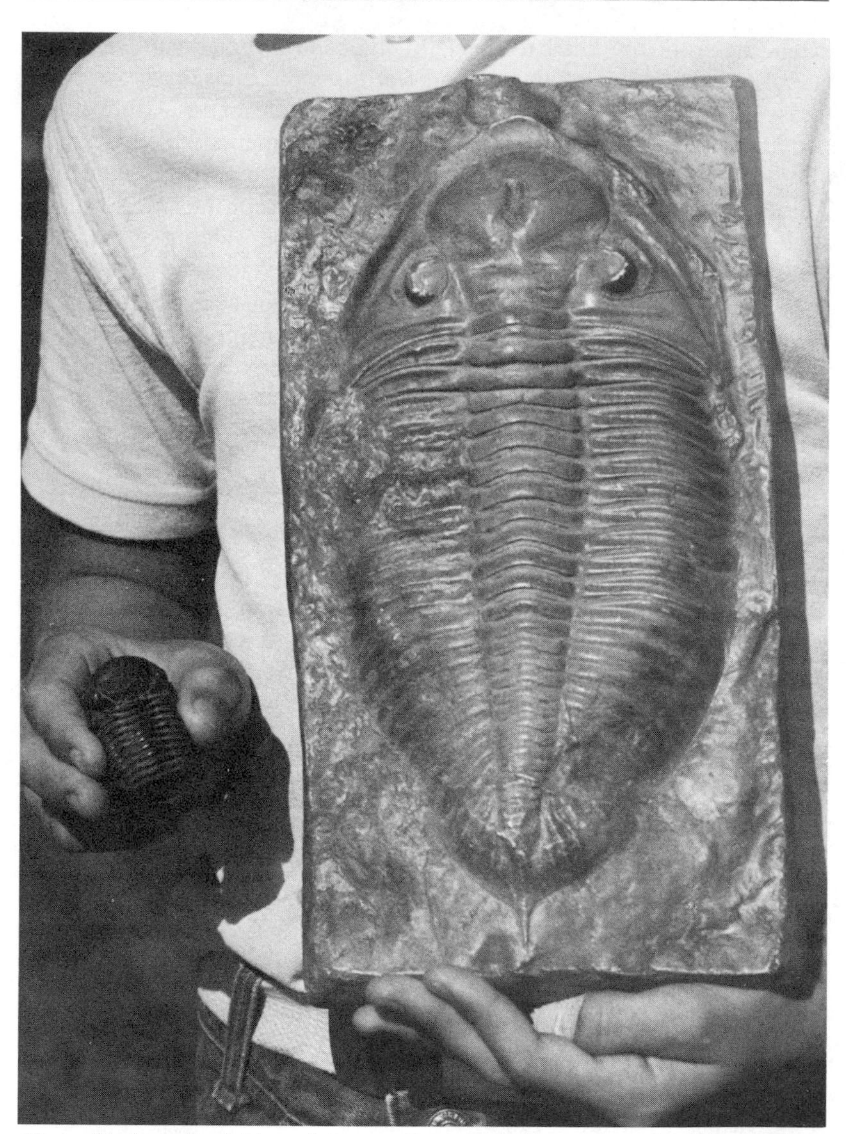

Figure 41. A small trilobite and the cast of a giant. What a gourmets' delight, if only we had time machines!

glabella uniting the free cheeks into a single plate. Although generally small, specimens may range from 1 to 20 centimeters in length.

Common genera are *Dalmanites* (Silurian to Early Devonian), *Calymene* (Silurian to Middle Devonian), and *Phacops* (Silurian to Devonian). The last two are often found in a curled position with the pygidium tucked under the head.

SUBPHYLUM CHELICERATA
CAMBRIAN TO HOLOCENE

These marine and land-living arthropods have a combined head and thorax (cephalothorax) and an abdomen which may or may not be fused to the other parts. Some have an elongated tail spike or telson. In front of the mouth are enlarged legs that have been modified into pincers (chelae) which give the group its name. The legs are restricted to the cephalothorax. None of the group has antennae. Living forms include the horseshoe crabs of the Atlantic Ocean, spiders, ticks, mites, and scorpions. Mites may be less than a half-millimeter in length, while some of the extinct eurypterids grew to 3 meters in length.

Class Merostomata
Cambrian to Holocene

These are the marine chelicerates with book gills. Instead of breathing through tubes in the body walls, there are paired openings leading into a series of flattened leaflike chambers where the oxygen–carbon dioxide exchange takes placc.

Subclass Xiphosura
Early Cambrian to Holocene

The horseshoe crab is the only survivor of this group. The cephalothorax is short and vaguely horseshoe-shaped. The abdominal segments are fused except in the primitive forms, but there are movable spines on the sides of the abdomen. There is a long, spikelike telson or tail. We include in this subclass the very primitive Aglaspida from the Cambrian and Ordovician which are sometimes separated into their own subclass.

Subclass Eurypterida
Ordovician to Permian

Commonly referred to as being scorpionlike, the head or prosoma includes the first six body segments. The legs are confined to the head, the first pair forming pincers that are usually short but do grow to large size in some genera. The second through fifth pairs of legs are generally walking legs, and the sixth pair are expanded into paddles. The head has a pair of small eyes near the midline and two large eyes near or on the lateral margin of the head. The remainder of the body is covered by a flexible, unfused exoskeleton; segments 7 through 12 bear lungs, 13 through 17 form the tail, and number 18 forms either an elongated spike (telson) or a rounded paddle.

Most eurypterids are between 10 and 20 centimeters in length, but one species of *Pterygotus* (Devonian) is 180 centimeters long.

Typical genera include:

- *Pterygotus* Ordovician to Devonian
- *Hughmilleria* Ordovician to Pennsylvanian
- *Eurypterus* Ordovician to Pennsylvanian

Class Arachnida
Silurian to Holocene

These are the air-breathing chelicerates which use either trachea or book lungs to take oxygen directly from the air. In this group are the familiar spiders, scorpions, ticks, and the tiny mites. In size they vary from mites less than a half-millimeter in length up to scorpions which sometimes reach about 17 centimeters.

Order Scorpionida
Silurian to Holocene

The scorpions are the oldest and most primitive members of the class Arachnida. The second pair of legs are enlarged into formidable pincers and the tail ends in a stinger which injects poison into prey or antagonists. *Palaeophonus* (Silurian) is the oldest-known scorpion.

Class Crustacea
Late Cambrian to Holocene

This group is dominantly marine, although there are land crabs and a few other terrestrial forms. They have gills either attached to the legs as in the trilobites or attached to the thorax near the legs. Some have no gills and breathe through the body covering. The body is divided into three parts: the head or cephalon, a thorax, and a pygidium or abdomen.

The three important subclasses of crustacea are the Ostracoda, small bivalves that are mainly marine: the Cirripedia or barnacles; and the Malacostraca, which include the crabs, lobsters, shrimp, prawns, and crayfish.

Subclass Ostracoda
Early Ordovician to Holocene

These are small (usually less than 5 millimeters in length) crustaceans that have a two-valved shell of calcite and chiton, hinged on the dorsal or top edge. They generally have a pair of lateral eyes and a single median eye. The legs are ordinarily reduced to seven pairs. The ostracods are far-ranging pelagic animals that evolve rapidly and make excellent guide fossils and are used as environmental indicators.

Genera include:

- *Ceratopis* Ordovician
- *Drepanella* Ordovician
- *Hollina* Devonian

Subclass Cirripedia
?Cambrian, Ordovician to Holocene

The common barnacle does not look like a crustacean, with its volcano-shaped calcareous house, but an examination of its internal structure and the development (ontogeny) of the individual show that it is indeed a perfectly good one. The free-swimming larvae go through several molts and then develop a bivalved shell like an ostracod. Soon they attach themselves to some solid object by their antennae and begin to build a calcareous exterior. The gooseneck barnacles grow a stalk which ends in a nut-shaped shell, while the acron barnacle lays down a base plate and then builds a tapered structure of six fixed plates and four movable plates that can close the opening at the top. The head is lost in this process, and the legs extend up to the opening to create water currents for carrying oxygenated water and food down to the animal's mouth.

Balanus (Eocene to Holocene) is the common barnacle of the seacoast. One species, *Balanus gregarius,* from the Late Miocene of the West Coast, reached a length of 30 centimeters or more. Near King City, California, in the Salinas Valley, they are so commonly plowed up in the fields that one resident has used them to build a stone wall in front of his home.

Subclass Malacostraca
Devonian to Holocene

This is the gourmet's subclass of arthropods. It includes the crabs, lobsters, crayfish, shrimps, prawns, and the inedible sowbugs or pill-bugs. All are marine except the land crabs, the freshwater crayfish, and the terrestrial sowbugs who live in moist places and eat the roots of your garden plants. All have the head and part of the thorax fused to form a cephalothorax. In crabs, the tail is tightly tucked under the cephalothorax, while in others, such as the prawns, it is used as a swimming oar.

The Malacostraca are variously divided among up to 10 or more orders. Of these orders the Mysidacea (prawns—Mississippian to Holocene), the Isopoda (pillbugs or sowbugs—Devonian to Holocene), and the Decapoda (crabs, lobsters, shrimps, and crayfish—Triassic to Holocene) are the most familiar.

Class Insecta
Middle Devonian to Holocene

The familiar insects are six-legged, two-winged arthropods with sharply defined head, thorax, and abdomen. Most insects are small, but there are living moths and beetles as large as small birds. *Meganeura* (Pennsylvanian) was the largest recorded insect. This dragonfly had a wingspread of 75 centimeters, or more than the length of the average person's arm.

The fossil record of the insects is amazing in that these generally fragile organisms are so commonly preserved in the fine sediments and under special conditions. In North America, insects are found in Pennsylvanian concretions from the coal beds and associated sediments at Mazon Creek, Illinois; from the early Permian Wellington Formation, a series of shales and limestones with some anhydrites in Kansas and Oklahoma; the early Eocene lake deposits of the Green River Formation in Wyoming; the Oligocene volcanic ash beds at Florissant, Colorado; and the Miocene lake beds in Stewart's Valley, Nevada.

PHYLUM ECHINODERMATA
?PRECAMBRIAN, CAMBRIAN TO HOLOCENE

If you have ever walked on ocean beach or explored the tide pools, you will have met the echinoderms with the spiny skins. The sand dollars, sea urchins, and starfish represent two of the three living subphyla of this group.

All echinoderms are marine, basically bottom dwellers with both attached and free-moving forms. A few are even pelagic swimmers. Initially they were bilaterally symmetrical, but then became asymmetrical, as can be seen in the larval forms whose development of the right and left become uneven. On this asymmetry is superimposed the five-way or pentameral radial symmetry seen in adults, but in adults a plane of bilateral symmetry can still be established through the mouth and the anus. The skeleton is internal and may be built either of fused plates or of free plates and ossicles formed of calcite. One of the striking features of the echinoderms is the water vascular system of conduits and sacs which is used for food capture, movement, and respiration. If you have ever watched an aquarium starfish move across the glass front of its tank you have seen its tube feet, the sacs of the water vascular system, in operation.

SUBPHYLUM CRINOZOA (PELMATOZOA) EARLY CAMBRIAN TO HOLOCENE

These are the attached echinoderms that have stems with holdfasts or direct attachments to the ocean bottom. Secondarily some have abandoned the attached life and become free-floating or swimming forms. The major part of the body or theca is cup- or nut-shaped. The theca may be attached directly to the bottom by holdfasts or there may be a stalk of calcite plates held together by a fleshy covering. They may have well-developed, flexible arms or fixed food grooves that channel food to the mouth by the movement of the tube feet. Only four of the nine classes are important in the fossil record. The other five are small aberrant or ancestral groups, generally of interest only to the specialist.

Class Edrioasteroidea
Early Cambrian to Mississippian

These Paleozoic echinoderms have flexible plates forming the flattened, circular body. The mouth is on the upper surface at the junction of the five curved or straight ambulacral grooves. The animals attach themselves to a solid object such as the shell of some other animal. The Ordovician genus *Carneyella* seemed to have been particularly attracted to *Rafinesquina,* an abundant, broad, thin brachiopod of the time.

Class Cystoidea
Early Ordovician to Late Devonian

These are echinoderms with or without stems that are highly variable in shape, with a body made up of irregular-shaped calcite plates which vary in number from about a dozen to over 200. The plates are pierced by numerous pores which might have served as part of their respiratory apparatus. There are no food grooves on the primitive forms, but the later kinds developed three, and finally five, food grooves. The five-sided symmetry is poorly developed and the food-gathering arms are variable in number.

Class Blastoidea
Silurian to Permian

These intriguing extinct animals are small, attached echinoderms with a body (theca or calyx) attached to the bottom by means of a jointed stem with holdfasts near the bottom. The calyx is a five-sided, nut-shaped calcite box with five ambulacral areas radiating down the side from the summit or top which has six openings: a mouth, an anus, and four spiracles for the exhausting of water taken in for breathing. Most blastoids are less than 25 millimeters in their greatest dimension (width or height), but some may be nearly 65 millimeters tall. They are particularly abundant and reach their peak during the Mississippian.

Representative genera include:

- *Pentremites* Mississippian to Pennsylvanian
- *Cryptoblastus* Mississippian
- *Timoroblastus* Permian
- *Globoblastus* Mississippian

Class Crinoidea
Ordovician to Holocene

These animals live in a cup-shaped theca or calyx which is surrounded on the upper edge by five flexible arms which may branch repeatedly. On the upper surface of the cup are the food grooves which extend the length of the upper surface of the arms, and the mouth and anus. The calyx may be fastened directly to the bottom by holdfasts on the base, or there may be stems similar to those of the blastoids. These stems may reach 20 centimeters in length. Some genera abandoned the tied-down mode of life and became free swimmers, propelling themselves with their arms.

In the past, as today, the crinoids were gregarious and grew in vast "gardens" on the sea floor. During the time of great abundance, as in the Mississippian, thick piles of loose plates from the stems and calyces built up on the sea floor forming extensive layers of limestone or reefs known as biostromes.

Typical genera include:

- *Platycrinites* Mississippian
- *Forbesiocrinus* Mississippian
- *Pentracrinus* Triassic to Jurassic
- *Uintacrinus* Late Cretaceous

SUBPHYLUM ASTEROZOA
EARLY ORDOVICIAN TO HOLOCENE

These are the starfish and brittle stars. All possess the basics of five or multiples of five movable arms, carrying tube feet on the underside. The mouth is in the center of the bottom surface and the anus is on the top. The skeleton is an internal arrangement of loose calcite ossicles, and there are scattered spines on the surface of the skin.

Class Asteroidea
Early Ordovician to Holocene

This class includes common starfish whose arms blend into the central body. The arms have open ambulacral grooves running from the mouth to the tips on the underside.

Because of the loosely connected plates that make up the skeleton, starfish usually break up when the flesh disintegrates. Complete starfish are rare in the fossil record. There are a few localities where they are common because of special conditions or mass burials. *Devonaster* from the Devonian was found in large numbers in New York State where they were apparently buried by a submarine mudslide. In the Jurassic Sundance Formation in Wyoming there is a locality where natural casts of small starfish about 5 centimeters in diameter have been preserved in fair numbers. Similar specimens have been called *Asteriacites* and listed as trace fossils caused by the resting marks of various sorts of starfish. The sediments were fine silts laid down on the bottom of the Late Jurassic Sundance Sea, but the cast-forming mechanism hasn't yet been worked out.

Typical genera include:

- *Devonaster* Devonian
- *Urasteralla* Ordovician to Pennsylvanian

Class Ophiuroidea
Silurian to Holocene

The brittle stars have slender arms that are distinct from the body. The ambulacral grooves are closed by a cover of calcite plates.

Typical genera include:

- *Onychaster* Mississippian
- *Ophura* ?Jurassic, Late Cretaceous, to Holocene

SUBPHYLUM ECHINOZOA (ELEUTHEROZOA)
EARLY CAMBRIAN TO HOLOCENE

These echinoderms usually have a solid fused skeleton, the shape basically spherical, but flattened in the sand dollars, or somewhat heart-shaped in some of the sea urchins. There are no arms, but the skeleton is usually covered with movable spines.

Class Echinoidea
Ordovician to Holocene

These are the sea urchins and the sand dollars. The tests are made of thin, fused calcareous plates. They range from globular to disk-shaped and when viewed from the top may be round, elongated, or heart-shaped. The mouth on the bottom surface is equipped with a five-toothed jaw device known as Aristotle's lantern. The anus may be on top, at the back edge, or on the bottom side near the back edge of the shell. The five-petaled ambulacral area is confined to the upper surface, but the tube feet and food grooves continue around the shell to the mouth. The surface is covered with spines which are moved by muscles in the skin.

Subclass Perischoechinoidea
Ordovician to Holocene

These are regular echinoids with varying columns of plates, from 1 or more in the interambulacral areas and from 2 to 20 in the ambulacral area. They have teeth, but do not have gill slits.

Typical genera include:

- *Diplocidaris* Early Jurassic to Early Cretaceous
- *Cidaris* Holocene

Subclass Euechinoidea
Late Triassic to Holocene

In this subclass there are two columns of plates in both the ambulacral and the interambulacral areas; the anus may be within or outside the apical area, and there may or may not be teeth and gill slits.

Typical genera include:

- *Hemiaster* Early Cretaceous to Holocene
- *Clypeaster* Late Eocene to Holocene
- *Scutella* Miocene to Holocene
- *Dendraster* Pliocene to Holocene
- *Encope* Miocene to Holocene

PHYLUM CHORDATA
MIDDLE CAMBRIAN TO HOLOCENE

The chordates are animals that have as common features the body divided into a head, trunk, and tail; a dorsal stiffening structure (notochord) present some time during their life cycle; gill slits leading from the pharynx to the outside; and a dorsal nervous system. In the case of vertebrates the notochord has been partially or wholly replaced by a column of bone or cartilaginous vertebrae.

SUBPHYLUM HEMICHORDATA MIDDLE CAMBRIAN TO HOLOCENE

These are primitive chordates with the notochord restricted to the anterior part of the head.

Class Graptolithina Middle Cambrian to Mississippian

These are colonial chordates that build a colony with a chitinous exoskeleton. After the larva establishes the first cup (the sicula), continued budding builds a string of cups or theca in single or double rows. These strings of individuals are called stipes; the colony, which may consist of several stipes, is called a rhabdosome. Graptolites had three distinct life styles. Some were attached to the bottom and were therefore sessile; some had a long strand or nema at the base of the colony which attached to floating seaweed; and some developed floats.

For a long time graptolites were known only from carbonaceous films on the shale partings, but since the late 1930s specimens have been found "in the round" in limestones and dolostones, providing a means for a more detailed study of their internal structures. Although the soft parts are unknown, enough has been determined from the theca to lead students to believe that we are dealing with hemichordates.

As graptolites are ordinarily found in dark shales in the absence of other fossils, it is concluded that the shales were laid down on quiet bottoms in a situation where there was no dissolved oxygen being circulated by water currents. Therefore, the "graptolite facies" or environment of preservation was not related to where the animals lived, but rather to where they came to rest after they died. There are some shale deposits with graptolites that do contain other organisms, but the associated faunas are impoverished. The bottom environment must have been a marginal one.

Order Dendroidea Middle Cambrian to Pennsylvanian

These are attached to the bottom by the base of the first theca (sicula). There are three types of thecae arranged in sets on the stipes. The colonies are leaflike with interconnecting dissepiments between the stipes.

Typical genera include:

- *Dictyonema* Late Cambrian to Mississippian
- *Dendrograptus* ?Middle Cambrian, Late Cambrian to Mississippian
- *Desmograptus* Ordovician to Mississippian

Order Graptoloidea
Early Ordovician to Late Silurian

This is the most important group of graptolites, pelagic forms that are either supported by a float (pneumatocyst) on the end of the nema or attached directly to some other floating object. The thecae usually are in one or two rows with only a few stipes in each rhabdosome. Because of their wide geographic range and rapid evolution, they are excellent index fossils. In spite of the fact that early paleontologists really didn't know what graptolites were, these fossils could still be of use in correlation because of their diversity, abundance, and restricted stratigraphic occurrence.

Typical genera include:

- *Tetragraptus* Early Ordovician
- *Diplograptus* Middle Ordovician
- *Monograptus* Early Silurian

SUBPHYLUM CONODONTOCHORDATA ?LATE CAMBRIAN, EARLY ORDOVICIAN TO LATE TRIASSIC, ?LATE CRETACEOUS

These tiny jawlike calcium phosphate fossils have been known since 1856 and have long been used as index or guide fossils. They range in size from less than 1 millimeter to 3 millimeters in length. Despite their widespread use as guide fossils and the description of scores of species, it was only during the past 10 years that the animal which bore conodonts was discovered in Montana.

Conodonts had been considered to be molluscan radulae, possibly from gastropods and cephalopods; the jaws of annelids, jaws, or claws of arthropods; and the jaws of a primitive fish.

It was known that they were grouped into sets, that there were right- and left-hand members of pairs, that they did not seem to be restricted to any particular bottom environment, and that they did not show signs of wear or abrasion. It was generally agreed that the conodont animal must be soft-bodied (as no definitely associated other hard parts have been found), bilaterally symmetrical because of the mirror image pairs, and marine and pelagic as conodonts were found only in marine sediments without regard to facies and in all parts of the world. When the conodont animal was found in 1968 these speculations turned out to be true.

So far nearly two dozen specimens of the conodont animal have been reported from rocks of either Late Mississippian or Early Pennsylvanian age in the Little Snowy Mountains of Montana. The animal is a fish-shaped, free-swimming chordate about 70 millimeters long. There is a dorsal fin fold for stability and a tail fin for swimming. Internally there is what appears to be a dorsal nerve chord and a notochordlike rod in the front quarter of the animal. The gut extends from an anterior mouth to an anus near the rear of the underside. The conodonts themselves appear in the large part of the gut in the middle of the body. They were disturbed in the preserved specimens, but appear to have been mounted in pairs, possibly in a vertical circle. Their inferred function was to form water currents through the gut by vibrating and to act as sieves.

As more is learned about the conodont animal it will be interesting to find out whether it is ancestral to the vertebrates or just a primitive side branch that lived for millions of years after more advanced forms had developed.

SUBPHYLUM VERTEBRATA (CRANIATA)
MIDDLE ORDOVICIAN TO HOLOCENE

These are the chordates with backbones and a cranium or brain box. Two materials form the skeletons of vertebrates: bone and cartilage. Initially they partially or completely replace the notochord, encase the brain, and then develop other skeletal features. Bone is the supporting tissue whose strength is derived from impregnation with phosphate and carbonate. Cartilage is a more flexible supporting tissue whose cells are imbedded within a chondrin matrix.

Superclass Pisces
Early Ordovician to Holocene

These are aquatic vertebrates with gills—all those animals that we broadly refer to as fish.

Class Agnatha
Middle Ordovician to Holocene

These are fish without jaws or paired fins. The gills are in a series of separate pouches that lead into the pharynx. They may have one or two nostrils and a third or pineal eye on the midline of the skull.

The earliest records of these primitive vertebrates are the scraps of body armor from the Ordovician. The Middle Ordovician Harding Sandstone in Colorado is a productive source for these scraps. The fragments tell us that bone had been developed and that these fragments are part of the body armor of primitive vertebrates known as ostracoderms (more particularly the Heterostraci). It is not until the Late Silurian and Early Devonian, when freshwater deposits became more common, that we find a good fossil record of these primitive armored fish.

Order Osteostraci (Cephalaspida)
Late Silurian to Middle Devonian

These are the classical ostracoderms, jawless vertebrates with flattened heads encased in a bone armor. Behind the head the body is covered with bony scales. There is a swimming tail in which the vertebrae tilt upward. The flattened head has a jawless mouth and rows of openings to the gill pouches on the underside. On the upper surface, set well back on the head, are paired eyes near the midline, a single medial pineal eye, and a single nostril. Around the side edges and on the midline are a series of organs that were once thought to be electric protective devices, but now are seen as water-pressure-sensitive balancing organs. The heads are variously shaped, some with a variety of "horns"projecting back from the rear corners and others with spikes projecting forward like rams. Some had paddles that may have acted as rather ineffectual swimming oars. The tail with upward-bent vertebral column and unsupported lower lobe (heterocercal condition) would tend to push the body down as the fish swam. This, in conjunction with the flat head and mouth at the lower side of the front of the head, suggests a bottom feeder who probably worked the mud for digestible tidbits. The average length of these fish was about 20 centimeters.

Typical genera include:

- *Cephalaspis* Late Silurian to Early Devonian, ?Middle Devonian
- *Hemicyclaspis* Early Devonian

Order Arthrodira
Late Silurian to Devonian

The arthrodires are certainly the most spectacular placoderms. *Dunkleosteus,* better but incorrectly known as *Dinichthys,* reached a length of 10 meters, and isolated fragments suggest that much larger individuals existed. In this order the head and chest armor are hinged on the sides with a ball-and-socket-type hinge formed by a projection on each side of the chest armor inserted into a notch at the back of the head armor. Some arthrodires lost much of their internal bone which was replaced by cartilage, and some lost the scale covering from the body behind the thoracic armor.

Typical genera include:

- *Dunkleosteus* Late Devonian
- *Coccosteus* Middle to Late Devonian

Order Antiarchi
Early to Late Devonian

These freshwater placoderms averaged about 15 centimeters in length. The head and thorax are covered with a bony armor, as in the arthrodires, but the two sections do not have the neat ball-and-socket joint arrangement. The posterior body and tail were covered with scales or naked. The body was flattened on the underside; the tail was heterocercal; and the eyes, all three of them, were clustered together on top of the head. All these characteristics suggest that they were bottom dwellers and feeders. Preserved soft parts indicate that these fish had lungs, a primitive characteristic of all the jawed fish except the sharks.

The most common antiarch is *Bothriolepis,* whose remains are abundant in the Late Devonian freshwater deposits along the shores of Chaleur Bay on the Gaspe Peninsula in Quebec. These fish grew to a length of about 30 centimeters and complete suits of their armor are to be seen in most museums.

Typical genera include:

- *Bothriolepis* Late Devonian
- *Plerichthyes* Middle Devonian

Order Acanthodii
Late Silurian to Permian

The acanthodians or spiny "sharks" are often placed with the higher fishes. Instead of a solid body armor on the head and thorax, they have a complete covering of diamond-shaped ganoid scales. These scales have a base of bony material and are covered with a thick coat of enamel. Occasionally there are bony plates in the gill region and surrounding the eyes. Most acanthodians are freshwater dwellers, but there are a few marine forms. They range from about 8 to 30 centimeters in length. The tail is heterocercal with all of the fin-forming lobes located below the vertebrae. The other fins have a bone spine on the leading edge to form a support. Some genera such as *Climatius* have more than the usual pectoral and pelvic pairs of lateral fins. These extra paired fins make a good case for the theory that fins were developed as folds of skin strengthened by bone spines, stabilizing the body as the fish propelled itself with its tail fin.

Acanthodians are the oldest-known placoderms and most closely resemble the later bony fish which have ossified internal skeletons and a covering of scales. They are specialized in ways that remove the known types from the direct ancestry of the higher fish, but they must be quite close to that unknown ancestral stock.

Typical genera include:

- *Climatius* Late Silurian to Early Devonian
- *Euthacanthus* Early Devonian

Class Chondrichthyes
Late Middle Devonian to Holocene

These fish represent the solution to an internal skeleton without the use of bone. Instead of bone, the skeleton is made of cartilage. The modern sharks, skates, and rays are the common members of the group, but there is also a little known subclass, the Holocephali or Chimaerans, which are wide-ranging omnivorous forms that prey on shelled invertebrates.

The sharks and their relatives differ from the other modern fish not only by having primary cartilaginous skeletons, but in having gills open to the outside through separate slits. The males have claspers on the pelvic fins to aid in internal fertilization of the eggs, and there are no lungs.

Although the sharks are predominantly marine, there are cases of land-locked forms adapting to freshwater for long periods of time. A good case is shown by the presence of rays in the Eocene Green River lake beds in Wyoming. These were survivors from the Cannonball Sea, a Paleocene remnant of the seaway that split North America from the Arctic Ocean to the Gulf of Mexico during the Late Cretaceous.

Order Selachii
Late Devonian to Holocene

These are the "modern" sharks with a large number of types spread throughout their long history. They have five to seven gill slits and a variety of tooth types, and some reach a length of 15 meters. Many of the genera and species are known only from teeth, as the cartilaginous skeletons are rarely preserved. Sometimes old individuals will have calcified the cartilaginous skeleton to the point where preservation is possible.

During the Mississippian, sharks with "pavement teeth" which covered the upper and lower jaws were very common. The low, rounded-tip teeth were developed to crush the shells of marine invertebrates. Most of the present-day sharks are active predators with pointed teeth for holding and tearing their prey. As these teeth are constantly being replaced it is possible for a single shark to leave a fossil record that suggests the presence of many individuals.

Common genera include:

- *Hybodus* ?Late Permian, Early Triassic to Paleocene
- *Hexanchus* Early Jurassic to Pleistocene
- *Isurus* Cretaceous to Holocene
- *Charcharodon* Paleocene to Holocene

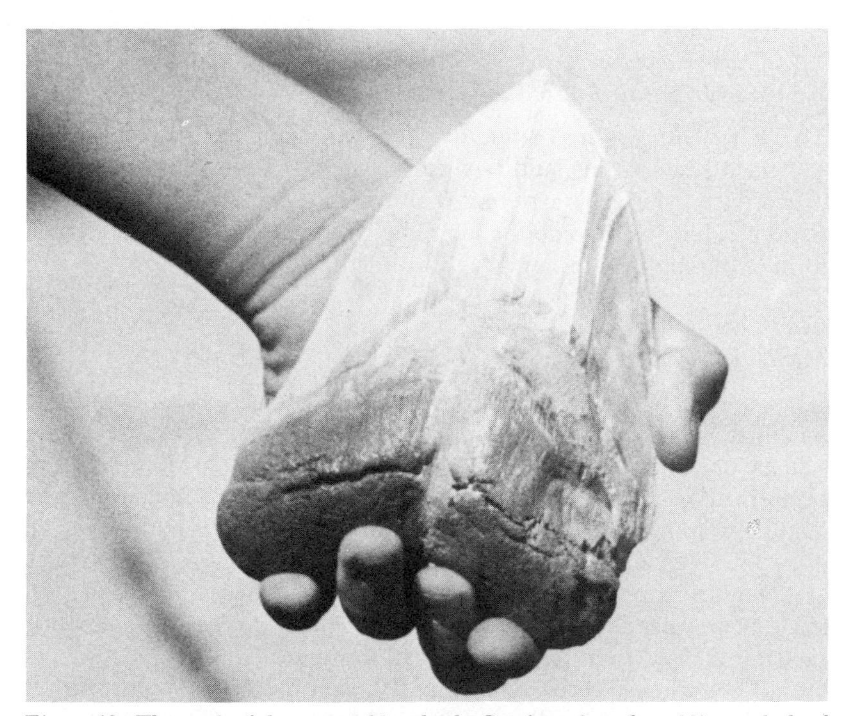

Figure 42. The tooth of the great white shark, Carcharodon, *from Miocene beds of the Calvert Formation along Chesapeake Bay.*

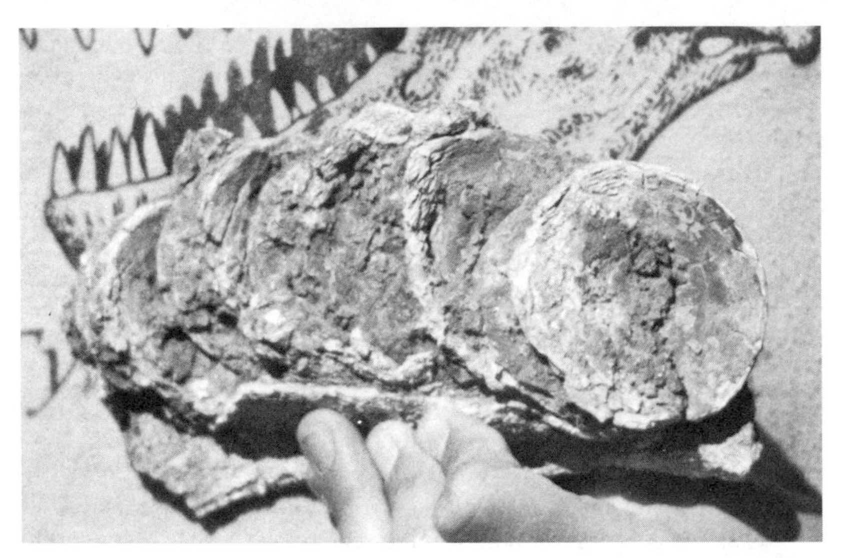

Figure 43. The calcified vertebrae of a large shark. As sharks' skeletons are made of cartilage, they are only preserved when the bones calcify during life.

Order Batoidea
Early Jurassic to Holocene

The skates and rays are bottom dwellers with flattened bodies, broadly expanded pectoral fins, and less well-developed pelvic fins. They feed primarily on shelled invertebrates and are equipped with broad, flattened crushing teeth to do the job. The bat ray *Myliobatis* is a common form living in many of our bays and inlets.

Class Osteichthyes
Middle Devonian to Holocene

These are the bony fish with well-ossified skeletons (unless secondarily modified), lungs or swim bladders, gills opening to the outside through a single opening covered by a movable bony plate, and scales that were primitive bone covered with cosmine (a substance very much like the dentine in our teeth) and an outer layer of enamel that may either be quite thick (ganoid scales) or very thin (cosmoid scales). These scales are found in primitive fish, while the modern ones have a very thin scale made of a wafer of bone. The head is completely encased in a bone skeleton derived from plates that form in the skin.

The earliest record of the bony fish is in freshwater deposits from the early part of the Middle Devonian. As they evolved they eventually moved into the sea, so that by the end of the Paleozoic Era they were the major group of marine vertebrates.

With over 30,000 living species occupying nearly every possible aquatic niche, including hot springs, the classification of the Osteichthyes is predictably complex. There are about 40 recognized orders spread through three subclasses and numerous infraclasses.

Subclass Actinoptergii
Middle Devonian to Holocene

These are the common modern fish with fins in which the bony supports form fanlike rays. The subclass is divided into three infraclasses: the primitive Chondrostei, the intermediate Holostei, and the advanced modern Teleostei.

Infraclass Chondrostei
Middle Devonian to Holocene

These are the primitive ray-finned fish that reached their peak in the Permian and then virtually disappeared by the beginning of the Cretaceous. Today the sturgeon, the paddlefish, and the bichir are the only survivors. These can usually be seen in any large aquarium. They have massive, ganoid scales and a number of primitive skeletal features. The lungs have not been modified into swim bladders.

Infraclass Holostei
Late Permian to Holocene

The holosteans are intermediate between the primitive and modern actinopterygians. This group reached its peak at about the Jurassic –Cretaceous transition. Most of them were gone by the end of the Cretaceous.

The garpike (*Lepidosteus*—Late Cretaceous to Holocene) and the bowfin (*Amia*—Late Cretaceous to Holocene) are surviving members of the group. In many freshwater Late Cretaceous to Eocene deposits the diamond-shaped scales of gars are common fossils. One interesting fossil genus is *Aspidorhynchus* (Middle Jurassic to Cretaceous), a freshwater fish about 60 centimeters long with tilelike scales and an upper jaw that extends into a pointed spear. Beautiful specimens, generally illegally collected in Brazil, are often found in rock and curio shops.

Infraclass Teleostei
Triassic to Holocene

The modern teleosts are at the peak of their development at the present time. Since the earliest record of *Leptolepis* in the Triassic, the group has been expanding until its members occupy virtually every available watery niche. The skeleton is fully ossified, the lungs have become completely modified into a swim bladder, and the scales are thin, nonenameled wafers of bone. Virtually every fish we encounter is a teleost.

SUPERCLASS TETRAPODA
LATE DEVONIAN TO HOLOCENE

These are the four-footed land vertebrates, some of which have become adapted to live in the water and others to flying.

Class Amphibia
Late Devonian to Holocene

The amphibians are the most primitive land vertebrates. The eggs are laid in water and the young go through a larval or tadpole stage. Breathing is by lungs or through the skin, which is glandular and without scales. The living frogs, toads, salamanders, and caecilians are divided into three orders. The Anura (frogs and toads) and the Urodela (salamanders and newts) are known to virtually everyone.

Order Temnospondyli
Late Mississippian to Late Triassic

These are the advance stegocephalians that flourished into the Triassic. They retained the solid-roofed head and labyrinthodont teeth. Most stayed close to the water, and many must have been almost completely aquatic. The heads were often broad and flat, with prominent grooves running around the top, remnants of the lateral line balancing system left over from their fish ancestors. The system was probably functional for the amphibians as a balancing aid when swimming in rivers or lakes.

Typical genera include:

- *Eryops* Early Permian
- *Metoposaurus* Late Triassic

Class Reptilia
Late Pennsylvanian to Holocene

The reptiles are the first complete land vertebrates.

Order Chelonia
?Middle, Late Triassic to Holocene

The turtles and tortoises developed their protective covering and then rapidly froze the design, remaining virtually unchanged since mid-Mesozoic times. They are found in marine, freshwater, and terrestrial environments.

Order Pelycosauria
Late Pennsylvanian to Early Permian

This group is not on the main line of mammal evolution, but it does have many mammalian characteristics. They have begun to reduce the number of phalanges to the 2-3-3-3-3 formula found in mammals, and the cheek teeth (those along the sides of the mouth) have different forms and functions.

The "sail lizards" *Sphenacodon* and *Dimetrodon* of the Early Permian and *Edaphosaurus* from the Late Pennsylvanian through the Early Permian typify the pelycosaurs.

Order Sauropterygia
Early Triassic to Late Cretaceous

The plesiosaurs are the most important members of this group. Their streamlined bodies and limbs modified into powerful paddles indicate completely pelagic life styles. The long-necked forms, known as the elasmosaurs, have barrellike bodies, a short steering tail, a long neck, and a very small head. They were fish eaters which cruised the seas of the world. Some reached a length of 17 or more meters. The marine Cretaceous deposits of Kansas, South Dakota, and California have produced a number of excellent skeletons that are displayed in natural history museums throughout the United States.

Order Ichthyosauria
Middle Triassic to Late Cretaceous

The ichthyosaurs were the whales and dolphins of the reptile world with their fusiform, fishlike bodies. The head had a long, pointed snout, large eyes, and a greatly shortened cranial region. The neck was greatly shortened as in the whales, the four limbs were small paddles, probably suitable only for balancing and steering, and the tail had developed into a powerful oar.

In North America ichthyosaurs are particularly abundant in the Triassic deposits of northern California and Nevada. In the latter state, near the town of Gabbs, is Ichthyosaur Paleontologic State Park, the scene of a mass stranding of giant ichthyosaurs almost 20 meters long.

Order Squamata
Late Triassic to Holocene

The lizards and the snakes are presently the successful members of this order. The lizards are quite common in many fossil faunas. Bony plates, jaw fragments, and vertebrae seem to be readily preserved wherever small land vertebrates are found.

During the Late Cretaceous one group returned to the sea to become the mosasaurs. Although not as fishlike as the ichthyosaurs, they did have limbs reduced to weak paddles and a powerful swimming tail. Some reached a length of more than 10 meters. Skeletons of these reptiles are often found in the same deposits where plesiosaurs are found. *Clidastes* from the Late Cretaceous is a spectacular mosasaur.

Order Crocodilia
Middle Triassic to Holocene

The crocodiles and the alligators are surviving primitive reptiles. They replaced the highly successful and related phytosaurs by the end of the Triassic and have held their tropical and subtropical river, lake, and swamp habitats ever since. They are distinguished by the development of a secondary bony palate as in the mammals. This separates the breathing and eating functions by a wall or floor of bone so the animal can eat and breathe successfully at the same time.

Order Pterosauria
Middle Jurassic to Late Cretaceous

Two reptilian groups become successful flyers. The birds were derived from small feathered dinosaurs which took to gliding as they ran. The origin of the pterodactyls is unknown, but it was surely different from that of birds. Pterodactyl wings are formed by a flap of skin extending from the tip of a greatly elongated fourth finger to the hind leg. One group is known to have had hair, so they may be presumed to have been warm-blooded. They were not flyers in a strict sense, but rather gliders who could probably stay aloft for long periods of time. As most are found in marine deposits with associated fish remains, their life style must have been such that nurture of the young was a necessity until they could fish for themselves. Recently nonmarine and non-fish-eating Cretaceous pterodactyls have been found in Texas. This indicates that there is another branch of the group which is unknown except for these few specimens.

Rhamphorhynchus of the Late Jurassic and *Pteranodon* from the Late Cretaceous are typical of their time.

Orders Saurischia and Ornithiscia
Late Triassic to Late Cretaceous

Together these two orders contain the reptiles known popularly as dinosaurs. They are certainly warm-blooded; some smaller ones must have had insulating feathers; and they were surely more active than generally pictured.

For years dinosaurs were pretty well ignored by paleontologists, except in spectacular exhibits in natural history museums.

Class Aves
Late Jurassic to Holocene

The birds have variously been described as glorified reptiles or even as living dinosaurs. They are warm-blooded, have feathers, and lay hard-shelled eggs. These are characteristics that may well have been shared with some extinct reptiles. They are unique in the skeletal modification accompanying their development of flight. The neck vertebrae have saddle-shaped articulations which give them a great deal of freedom of movement while retaining a strong joint. The body vertebrae are fused into a single unit providing rigid wing support. The "ankle" and "wrist" bones have been fused to the neighboring long bones, which have, in turn, been fused together where there are two or more paralleling each other.

Class Mammalia
Triassic to Holocene

The mammals combine a number of unique characteristics obvious in the living mammals but not observable in the fossil forms. Important skeletal characteristics often seen in fossil remains are a single bone (the dentary) on each side of the lower jaw, which articulates with the squamosal in the cranium; differentiated teeth; the transformation of the reptilian articular and quadrate bones into the malleus and incus of the ear; a double occipital condyle or joint between the cranium and the neck vertebrae; and a basic phalangeal formula 2-3-3-3-3. In recent years the presence of a single bone in the lower jaw has become less important as a defining characteristic of the mammals, and animals with extra jaw elements are admitted to the class, providing the dentary–squamosal articulation has been achieved.

Chapter Six

LET'S LOOK AT THE UNITED STATES FOR FOSSIL AREAS

Space isn't great enough here for a complete rundown, but I'll give you some ideas. There are a number of books available with "detailed" directions to fossil-bearing localities, but they offer more than they can provide. One is J. E. Ransom's *Fossils in America*. It contains a state-by-state description of where fossils may be found. Unfortunately, it is badly flawed.

Many of Ransom's localities are strictly off limits. For example, he cites Rancho la Brea in Los Angeles. How long would you last in this county park if you started digging up the lawns? He cites Comanche Point of the south end of California's San Joaquin Valley. This is in a large ranch holding where I had to go through two levels of management to get permission to take a class for a day's prospecting.

In one part of his book, Ransom states that Agate, Nebraska—the site of Agate Fossil Beds National Monument—is in Nebraska. In another place, Sioux County, the home of Agate, is in South Dakota.

One thing to remember is that the U.S. Antiquities Act covers collecting on federal land. Before collecting on federal land check with the local management, usually the Bureau of Land Management, the Forest Service, or the Army Corps of Engineers. In general, collecting plants and invertebrates is permitted, but fossil vertebrates are off limits.

Each state has individual laws regarding public lands and trespass. Be sure to check these before going into the field. Always get permission to work on private land.

You will have to find your own way through the country's geology to locate possible fossil prospecting areas. Sedimentary rocks are everywhere; all you have to do is find the kind and age of sediments that may contain the sort of fossils you are interested in.

The U.S. Geological Survey has a four-sheet geologic map of the "Lower Forty-eight" which will give you the big picture of the country's geology. In addition, the Survey has published a large number of geologic quadrangle maps of scattered, generally economically important areas.

Many state surveys have also done extensive geologic mapping. For example, the entire state of California has been mapped geologically to the scale of 1:250,000.

The American Association of Petroleum Geologists (AAPG) has published a series of excellent regional geological highway maps. These are:

1. Mid-Continent Region (Kansas, Oklahoma, Missouri, and Arkansas).
2. Southern Rocky Mountain Region (Utah, Arizona, Colorado, and New Mexico).
3. Pacific Southwest Region (California and Nevada).
4. Mid-Atlantic Region (West Virginia, Delaware, Maryland, Kentucky, Virginia, Tennessee, North Carolina, and South Carolina).
5. Northern Rocky Mountain Region (Idaho, Montana, and Wyoming).
6. Pacific Northwest Region (Washington and Oregon).
7. Texas.
8. Alaska–Hawaii.
9. Southeastern Region (Louisiana, Alabama, Georgia, and Florida).
10. Northeastern Region (New York, Vermont, New Hampshire, Maine, Massachusetts, Connecticut, Rhode Island, Pennsylvania, and New Jersey).
11. Great Lakes Region (Wisconsin, Michigan, Illinois, Indiana, and Ohio).
12. Northern Great Plains Region (North Dakota, Minnesota, South Dakota, Nebraska, and Iowa).

These excellent maps are very instructive. In addition to the geologic map of each region are geologic columns, geologic cross sections, a series of small maps showing the historical geology of the region, a tectonic map, and in some a satellite picture of the area. All of this is backed up by a very readable text.

I suggest for starters that you write for a map of the area you are interested in:

American Association of Petroleum Geologists
P.O. Box 979
Tulsa, Oklahoma 74101

A little diligence will bring other good geologic maps to the surface. The Arizona Geological Society, P.O. Box 4489, University Station, Tucson, Arizona 85717 has published the "Arizona Highway Geologic Map." This includes a geologic map, a historical geology review, and other good information.

Utah and Texas have also produced statewide highway maps. The Utah map "Geological Highway Map of Utah" was produced by the Department of Geology, Brigham Young University, Provo, Utah 84602.

These maps are your key to searching for fossils. Use them.

Contact your state geologic survey, state bureau of mines, and the geology departments of local colleges and universities. They can either tell you what is available or direct you to a map source.

The Pacific Northwest Region (AAPG Geological Highway Map Number 6) comprises Washington and Oregon. These two states are largely covered with volcanic rocks, with the Cascade Mountains in the west and the Columbia River basalts to the east. However, there are fossil-bearing Cenozoic deposits along the Pacific Coast. The Oregon coast has produced a great number of spectacular marine mammals, particularly whales, seals, sea lions, and the weird hippopotamuslike bottom walkers, the desmosoylids. Inland there are widely scattered outcrops of Tertiary continental deposits which, with a great deal of searching, will yield fossil mammals. Any of these should be shown to a paleontologist as they could be scientifically important. There are even scattered outcrops of Mesozoic and Paleozoic marine sediments which could produce interesting specimens.

The Pacific Southwest Region (AAPG Geological Highway Map Number 3) is geologically a very diverse area. The Coast Ranges north of San Francisco Bay are made up largely of metamorphic rocks, dominated by the Franciscan Formation and scattered volcanics. Inland, the north is primarily covered with volcanics. North of Shasta Lake there is a large area of Paleozoic and Triassic marine deposits. These are very fossiliferous in spots.

The entire eastern side of the northern and southern Coast Ranges are made up of Cretaceous marine sediments and a mixture of continental and marine Tertiary sediments. The Great Valley is generally unfossiliferous as are much of the Sierra Nevada, which are mainly composed of granitic intrusives and metamorphosed Paleozoic rocks. There are patches of continental Tertiary on the western flank at higher elevations, but below the glaciated zone.

The Transverse Ranges are a mixture of Tertiary beds which are often fossil-bearing. The continental sediments are particularly interesting as the rare fossils are scientifically important.

South of the Los Angeles Basin are the Peninsula Ranges, which

Figure 44. This is not an infantry entrenchment but a trench dug by fossil collectors on the bone level at Sharktooth Hill near Bakersfield, California.

are mainly granitic intrusives, but there is a narrow band of Cenozoic deposits along the Pacific Coast and scattered outcrops within the mountains.

To the east of the California portion of the Cascades, the Sierra Nevada, and the Penisula Ranges is the Basin and Range Province. This comprises all of Nevada, a good portion of Southern California and fringes of Oregon, Idaho, Utah, and Arizona. Someone once said that looking at a map of Nevada is like looking at a herd of caterpillars crawling slightly to the east of north.

The ranges are made up of Paleozoic and Mesozoic sediments, some Mesozoic intrusives to the south, and Tertiary volcanics. The floors of the basins are mostly made up of Quaternary alluvium with some tertiary sediments peeking through. The pre-Tertiary rocks yield a variety of invertebrates and ichthyosaurs. The following is an article, "Here There Be Sea Monsters," reprinted from *Gems and Minerals Magazine:*

> Look at the ichthyosaur bones partially freed from their entombing rocks at Nevada's Ichthyosaur State Park. Stand beside Sculptor Hoff's great reconstruction of a living ichthyosaur, shin-

Figure 45. A few minutes digging produced these shark teeth and marine mammal bone fragments at Sharktooth Hill.

ing in the bright sun and dry air of the desert. These two experiences should begin to give you a sense of the immense and complex history of the earth and its inhabitants.

What are the bones of a sea-going reptile, who lived 160 million years ago, doing in the rocks of West Union Canyon, Nevada, a site hundreds of miles from the ocean today and many feet above sea level? As far as that goes, what was a giant reptile doing living in the sea?

Let's consider the problem of the bones in the rocks first. During much of the decipherable geologic history of North America the continent was quite different from the one we know today. Where the Appalachian and Rocky Mountains rise high in the air, were once great shallow troughs often flooded by the sea. Between the troughs was a wide lowland, parts of which slowly and erratically sank into the sea and then rose like giant elevators. Sometimes the central continent was completely covered by the water, sometimes completely above it, sometimes sending up sea-girt mountains now long gone and destroyed by erosion.

The Appalachian trough disappeared when it was pushed up in

a series of folded mountains about 230 million years ago, just before the ichthyosaurs or "fish-lizards" appeared in the fossil record. The Rocky Mountain trough lasted until about 70 million years ago when it too was destroyed by mountain building, near the end of the Age of Dinosaurs. At that time the last of the great shallow seas that covered the continent slowly began to withdraw and the land as we know it took form.

While it existed, the western or Rocky Mountain trough was sometimes met by deep bays that extended from where the Pacific coast is today, eastward nearly to where the Rocky Mountains are now. During the Age of Reptiles a long wrinkle in the earth's crust separated these western embayments from the interior seaway. This broad, low wrinkle, trending north and south, has a really jawbreaking name given it by geologists. They call it the Mesocordilleran Geanticline. That just means it was a giant upwarp in the center of what is now our western mountain belt.

It is on the western shore of this wrinkle that the ichthyosaurs were stranded. It doesn't seem likely that they could have been chased ashore by enemies or stranded by the tide. Most paleontologists think they were stranded by a storm or died at sea of natural causes. If they died natural deaths the carcasses would have been buoyed up by the air in the lungs and perhaps carried ashore by winds and currents.

Sediments carried off the wrinkle by streams and thrown back up on the beach by waves buried the carcasses. Eventually these sediments were compressed and cemented into rocks, and the minerals carried by subsurface water petrified the bones so that they did not break up and disappear.

Beginning near the end of the Age of Dinosaurs, folding and uplift of the Rocky Mountains slowly lifted the old seashore deposits thousands of feet above sea level. Immediately they were attacked by the processes of erosion which began to tear them down again. After millions of years of burial the now-petrified bones of the ichthyosaurs were once again exposed to the air, this time not along the edge of an ancient sea but high in the dry desert mountains. If the processes of erosion continue long enough, and the mountains are not further uplifted, the bones will eventually be worn away like the rocks and, after millions of years, this area will once again be back at sea level.

Briefly, this is the sequence of geologic events which set the stage for the exhibit we can see today. Further and more complete information is readily available in any textbook on historical geology. The geologic history of North America, and particularly the western Cordillera—the mountainous region extending from the Rockies to the Pacific Coast, north to Alaska and south into Mexico and then to South America—is well worth reading.

Our second question asked what any self-respecting reptile had in mind when he left the land to imitate the fish. The answer can be given in two parts. One, all life seeks constantly to fill all the living space available to it. And two, every group of animals, under pressure, expands its range to include every ecologic niche it can fill.

Reptiles were the first well-adapted land animals with backbones. They were preceded (by millions of years) by their ancestors the amphibians. However, amphibians had to (and still must) lay their eggs in fresh water, so it can't really be said they had conquered the land. Some amphibians evolved into reptiles when they developed a self-contained egg, a living "space capsule" with a complete life support system to nourish and support the developing embryo until it was a small replica of its parents. Reptiles are born fully formed, not as fishlike tadpoles which hatch as partially developed embryos and must then face the hazards of the open pond while metamorphosing into adult form.

But why would reptiles re-invade the water when they had successfully conquered the land? Population pressure is always the answer. We can see land reptiles evolving into future marine reptiles in the Galapagos Islands today. The Galapagos Marine Iguana is slowly becoming a complete sea-going lizard. The other land iguanas are too numerous and food is too scarce. Though it still breathes with its lungs and lays its eggs on land, this species is potentially the mosasaur, mesosaur, or even ichthyosaur of the future.

When there isn't enough food on land to go around, some individuals in a population may try the water as a hunting ground. Perhaps at first they only scurry along the shore where they might catch unfamiliar things to eat. Later, as beachcombing becomes a way of life, they may expand their activities and chase things into the water. Ultimately if there is enough possibility for change built into the population's genetic makeup, and if natural selection does its thing, a new type of animal develops. He is no longer a land dweller but a well-adapted ocean ranger whose final form will be limited only by the available genes and the available time.

Such must have been the early history of the ichthyosaurs. To date we are still missing the early Triassic transitional forms which would link the ichthyosaurs with their Permian reptile ancestors. This does not worry paleontologists because link forms are constantly being discovered, the missing links seldom offer many surprises to the paleontologist, as the links can usually be pretty well imagined. But their presence is studiously ignored by the anti-evolutionist who constantly harps on their absence.

Ichthyosaurs are known for marine deposits of Mesozoic age all over the world. Those first recorded from the middle Triassic had essentially completed the evolution of the group. For the rest of the ichthyosaurs' existence, some 150 million years, there were only slow model changes and improvements.

Ichthyosaurs got their name (Greek for "fish-lizard") from their streamlined fishlike bodies. The beak was long and slender, with up to 200 conical teeth. The primitive Triassic forms held their teeth in sockets, but later species had long grooves the length of their jaws which held the teeth. Behind the snout the brain case ballooned, holding a pair of very large eyes protected by a ring of loose bony plates like the iris of a camera. A third smaller eye located in the forehead was a holdover from their primitive reptilian, amphibian, and fishy ancestors.

Behind the head stretched the vertebral column, up to 200 vertebrae. As these no longer had to support the body weight on land they had become simple disks without complex interlocking devices. The body was so streamlined that the head blended into it without any apparent neck. The double-headed ribs attached to the sides of the vertebrae along more than half the length of the body. Behind the ribs was a stout tail used to propel the ichthyosaur through the water.

Near its end the tail vertebrae angled downward as though broken. When completely preserved skeletons were first found it was believed that a tightening of the tail muscles and tendons after death had indeed broken the tail. Later, complete skeletons, the body outline preserved as a carbonaceous film on finegrained sediments, were discovered. These showed that in life the Triassic forms had a single-lobed tail with the vertebral column bent down to support and strengthen it. There was no corresponding upper lobe, just a fleshy frill. By late Jurassic time the tails were double-lobed like a fish's but still the vertebral support was in the lower lobe. The preserved body outlines also showed a triangular dorsal fin in the middle of the back, a sort of keel on top to help maintain an upright position in the water while swimming, without too much roll. We don't know if this dorsal fin had any support besides muscles or not.

The limbs had been modified into paddles for steering and balancing in the water. In some cases the number of fingers and toes had been increased to make the paddle wider, and the number of joints had increased to give more length with flexibility. Other groups added joints without adding extra fingers and toes, and even developed a long, slender, pointed flipper by losing one digit altogether.

Ichthyosaurs are the most highly adapted aquatic reptiles known. Like their modern counterparts the mammalian whales, dolphins, and porpoises (whom they greatly resemble), they must have lived only in the water throughout their lives, coming to the surface to gulp air into their lungs and then diving again.

We know that the young were born alive. That is, the eggs were retained in the mother's body until they hatched. The skeletons of some two dozen females with young inside have been found. One skeleton was remarkably preserved with a partially born baby caught in the pelvic canal and the tail and the rear half of the body outside, the head and the rest of the body still within the mother's body.

Ranging through the Mesozoic seas, the carnivorous ichthyosaurs probably hunted anything that looked edible (and wasn't bigger than they were). Preserved fragments from the stomach areas show that fish and squids were favorite items on the menu. But surely anything they could overtake and ingest was part of their diets.

Most ichthyosaurs were perhaps six or seven feet long, so the Nevada specimens were certainly giants. A fifty-six footer was a respectable size, but seventy-five foot specimens have been found. Because they lived during the Triassic Period they probably did not have a fully developed upper lobe on their tails. So the tail, with just an upper frill as depicted in Sculptor Huff's fine bas relief, shows their probable silhouette. Their flippers were long and narrow. Being among the more primitive ichthyosaurs, they retained recognizable upper and lower leg bones which in later forms became shortened and very much like the wrist and ankle bones.

Here in the mountains of Nevada, then, we have a truly remarkable animal preserved in the rocks, a triumph of evolution which, through a process we call adaptive radiation, sought to fill an improbable but possible ecologic niche. He is another of many fossil wonders of the ancient past.

Ichthyosaur Paleontologic State Monument is located northeast of Gabbs, Nevada, which is east of Walker Lake. Take State Highway 23 north (from Gabbs) for 1 mile, then turn east on a new paved road and drive for 22 miles through the hills. Take a graded gravel road for two miles to the monument headquarters. The old mining town of Berlin is part of the monument and the mill is being partially restored. The monument is open all year. Picnic spots are available as well as 14 campsites and drinking water.

The Cenozoic continental deposits are not very fossiliferous, so what is found is scientifically important.

The Northern Rocky Mountain Region (AAPG Geological Highway Map Number 5). This region includes Idaho, Montana, and Wyoming. This probably is the most diverse geologic area in the United States.

Idaho is mostly covered with volcanic rocks, but there are fossiliferous areas. In southwestern Idaho along drainages of the Snake and Boise Rivers there are late Tertiary lake, riverene, and flood plain deposits that are overall barren, but locally rich in fossils. In the Payette Formation above Boise, people are constantly finding "monkey" jaws. Actually the jaws are of a shell-smashing carplike fish. Farther east on the Snake River is the Glenn's Ferry formation which is loaded with freshwater invertebrates and fish in some areas. It also has the world-

Figure 46. Cracking nodules on the Snake River in Idaho. See Figure 4.

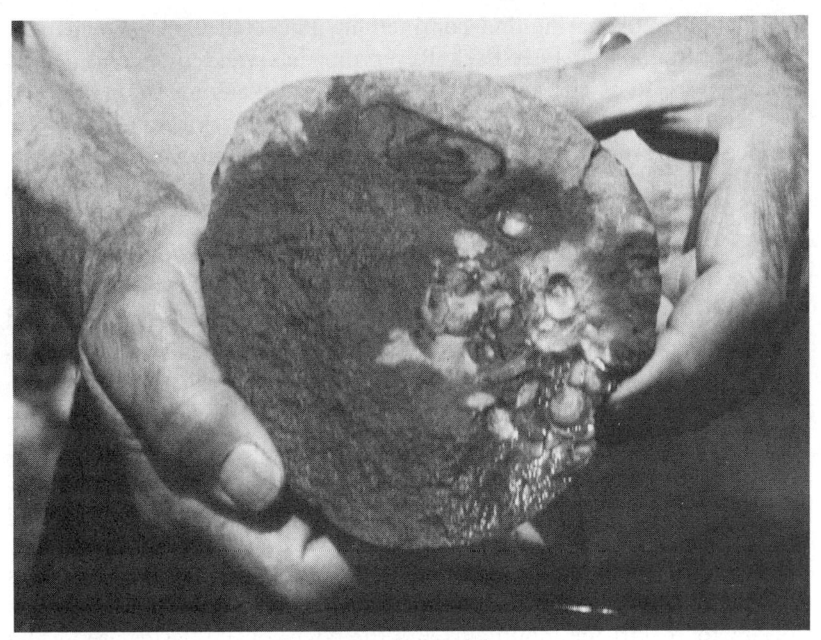

Figure 47. Freshwater snails in a cracked nodule.

famous Hagerman Horse Quarry (a no-no—it's on a school section); hundreds of late Pliocene horse skulls and dozens of skeletons have been recovered here during the past 50 years. In addition there have been many other types of mammals and birds collected. There are miles of exposures, but again, check on who owns the land, see if you are welcome.

Idaho from the southern base of the panhandle east is an area of folded and faulted mountains. Here are sediments ranging from Precambrian to Cretaceous in age with Tertiary sediments in the intermountain basins.

Northeastern Idaho and northwestern Montana have hundreds of square miles of the Precambrian Belt Series exposed on the surface. These beds have algal masses up to a yard in diameter. These are beds that should be thoroughly prospected for other Precambrian fossils. Someone, someday will make an international reputation by discovering a good fauna of soft-bodied animals in these beds. In the southwest and eastward across the center of the state are a series of folded mountains built of folded sediments ranging in age from Cambrian to late Mesozoic. In many places these rocks are rich in invertebrates. The Jurassic rocks are particularly notable for the five-pointed star-shaped column plates of the crinoid *Pentacrinus*.

To the north and south are great expanses of flat-lying Cretaceous sediments, both marine and continental. The Hell Creek Formation, exposed around the Fort Peck Reservoir, has yielded several *Tyrannosaurus* skeletons and many skulls and a few skeletons of *Triceratops*.

The eastern edge of Montana has Tertiary rocks exposed on the surface with a few tongues of Cretaceous sediments reaching the surface. The Tertiary rocks range from Paleocene to Pliocene in age and do produce fossils.

Wyoming is another state that has something for everybody. It is a land of dome and fault mountains with intermountain basins with floors of Paleocene and Eocene sediments. Most of these are floodplains and channel deposits, but in the southwest are the Eocene lake deposits which yield the beautifully preserved Green River fish.

The northwestern corner is mostly volcanic and part of Yellowstone National Park. Here the layers of standing petrified forest are preserved in volcanic ash. A no-no, of course. The rest of the state is dominated by mountains and basins. The Tetons, Absaroka, Wind River, Big Horn, Casper, Laramie, Medicine Bow, and the northern edge of the Uinta Mountains dominate the landscape. In these are rocks ranging in age from the Precambrian to Cretaceous. The Cretaceous rocks which skirt the mountains and the edges of the basins are locally rich in ammonites. The prominent cliff-forming Madison Limestone of Mississippian age is a good source of marine invertebrates. The Mississippian is the time when extensive massive limestones were deposited in the Rocky Mountain area. These limestones have various names. The common ones include the Madison of Wyoming, the Pahasapa of the Black Hills of South Dakota, and the Redwall in the Southwest.

The Tertiary rocks of the basins are mainly of Paleocene and Eocene age. The Big Horn Basin has extensive outcrops where the surface collecting of Eohippus teeth and parts of other early Eocene animals is not difficult. Check with the Bureau of Land Management in Worland for the current ground rules. As mentioned above, the Eocene lake beds in southwestern Wyoming are the homes of the famous Green River fish. Here, a visit to Fossil Butte National Park and Carl Ulrich's studio and fossil laboratory at Fossil, Wyoming a few miles west of Kemmerer, Wyoming are a must. There are several privately owned quarries in the area where for a fee you may do a day's collecting.

Here is an article from *Gems and Minerals Magazine*, "Fishing in the Great Lakes of Wyoming":

> The background for the Great Lakes of Wyoming begins at the end of the Cretaceous Period (the last of the "Age of Dinosaurs") when the Rocky Mountains began to be uplifted, folded and faulted, forming a series of down-folds or structural basins surrounded

by folded mountains in the area we now know as Colorado, Wyoming, and Utah. As the mountains rose they were rapidly eroded in the wet and semi-tropical climate of the time. Rivers running down from the heights carried sediments which slowly filled the basins within the mountain chain. These were the sediments which also entombed many of the animals living during those early days of the "Age of Mammals" so we have there a wonderful record of the beginnings of our own geological era, the Cenozoic.

Several of the structural basins became the sites of huge lakes which persisted for millions of years. They expanded and contracted in response to the changing climatic cycles. The earliest of these lakes was Lake Flagstaff in Utah, extending from just south of Provo to Bryce Canyon during the late Paleocene, the first epoch of the Cenozoic. Lake Flagstaff slowly disappeared and was replaced by Uinta Lake during the early part of the Eocene Epoch. Lying south of the Uinta Mountain in Utah, this lake extended from Provo on the north and west to Meeker, Colorado, on the east and to Grand Junction, Colorado on the southeast. (While we are not exactly concerned with Lake Uinta in this story, it may prove to be very important to our futures. It is in these lake beds that the great oil reserves in the form of oil shales are stored. In lakes, sediments accumulate millimeter by millimeter in yearly strata called varves. Varves consist of two layers, one of clay deposited during the summer and a layer of carbonaceous material formed during the winter from the remains of algae, plants, and animals that sink to the bottom during that season. In quiet deep water an organic ooze forms which is later compacted into a material called kerogen. Under ideal conditions this accumulation will yield sixty or more gallons of crude oil for each ton of shale processed.)

North of the Uinta Mountains in Wyoming the Green River lakes developed in the middle Eocene. The Green River basin covered much of southwestern Wyoming. This great basin was itself divided into two basins, the Bridger Basin on the west and the Washakie Basin on the east. The history of these various lakes connecting and disconnecting has not been completely worked out. Geologists call the major body of water Gosiute Lake, but that was only one of the lakes in the Green River system.

In those days the climate of Wyoming was subtropical; warm rainy conditions in the late spring and early summer followed by a dry, hot late summer and early fall. Winters were mild. Under these conditions coal formed in some areas; in others, where the basins were subjected to flooding with mineral-rich waters and then evaporated during the dry season, thick deposits of trona, a hydrous carbonate of sodium which is extensively mined today, were left behind. As in Uinta Lake, there are also extensive deposits of oil shale in these lake beds.

To us, however, the most interesting part of the Great Lakes deposits are the millions of fish preserved in the fine-grained lake bottom sediments. The first recorded discovery of these fish fossils was in 1852. Dr. David Dale Owen, who was making a geological survey of Minnesota, sent his assistant, John Evans, westward to explore the Mauvaisses Terre or the White River Badlands of South Dakota, as they are called today. Evans went on into Wyoming and collected fossil fish from somewhere in the Green River lake beds. His specimens ended up in Philadelphia where Dr. Joseph Leidy, considered the father of vertebrate paleontology in North America, described the material in 1856. He named a new species of herring from the lot, but all we know of the locality is that it lies "on the Green River." Such a locality designation was common a hundred years ago when a third of our country was unknown territory to the average person. Scientists of that day would have scorned the nit-picking ways of today's paleontologist with his insistence on precise geographical and geological data.

When the Union Pacific Railroad laid track along the south side of the Green River in the 1860's, three miles west of the present town of Green River was a cut so full of fossils that it was named "Petrified Fish Cut." There many fish were collected when the excavation was made in 1868. This is also the place where Major J. W. Powell unloaded his equipment from the train and began his epic journey down the Green and Colorado Rivers in 1869. Since 1870 most of the fish fossils were taken from what was believed to be the Twin Creek locality. Later known as Fish Cliffs, this site is now Fossil Butte National Monument. Located just a few miles west of Kemmerer on U.S. 30 North, and just north of Fossil, Wyoming on the Union Pacific's Oregon Short Line, the site was worked commercially for many years. Beautiful fish were peddled at gas stations and curio stands along the railroad and the highway. If memory serves, the going price was about a dollar an inch.

When we visited the locality in 1937 commercial quarrying was being carried on by Mr. Haddenham of Kemmerer who had worked the site for some 40 years. The main fish bed was about a foot thick, divided by a thin bed of oil shale. The fish are in a moderately hard calcareous mudstone or marl that splits easily along the bedding planes. A quarry site is chosen where the fish layer is jointed with vertical cracks at two to four foot intervals. The overburden is blasted and individual blocks are pried loose and mounted on saw horses for splitting. Long, slender chisels were used to halve each slab, and the halving was repeated as many times as possible. If there are fish in the block, they lie between bedding planes, so they can be located by ripples in the thin matrix of limy clay that covers them. The string of vertebrae is the most obvious clue to a fish skeleton. When fish are found, the thin slab is sawed into blocks and the fish are prepared for sale or display.

The most common fish in the Great Lakes of Wyoming are the herrings, *Knightia* and *Diplomystus,* with the bass, *Priscacara* in third place. *Knightia,* usually just a few inches in length, is often found massed together, and very fine slabs of *Knightia* schools are fairly common. *Diplomystus* may reach 20 inches in length. The bass or "sunfish" *Priscacara* were the most popular items when the fossils were being peddled from the train station platform at Green River, and indeed they do make handsome specimens.

More rarely found and therefore more highly prized and priced were gar, paddlefish, and bowfins. Also in the rarity category are the stingray. Ordinarily salt water animals, the skeletons from Wyoming represent landlocked survivors left behind when the great Cretaceous seaway that split North America from the Arctic Ocean to the Gulf of Mexico retreated at the end of the "Age of Dinosaurs." Real rarities, and always of scientific importance, are the occasional skeletons of birds, bats, turtles, and even crocodiles that the lucky prospector occasionally finds.

The many large accumulations of herring skeletons suggests times of mass death. We still don't know what caused them. Suggestions have been made that there were "red tides" in which the waters were poisoned by growths of algae, or even that some of the thick falls of ash from volcanoes in the Yellowstone or Black Hills region could have caused suffocation. Whatever the cause, these intriguing occurrences have made the Green River beds one of the world's best known fossil-bearing formations.

With the establishment of Fossil Butte National Monument, the old quarry sites should someday be on display, and you may be able to see outstanding specimens in an accompanying museum. It may even be that Fossil, Wyoming, never quite the same since the railroad abandoned its grade through town, will rise again as a first-rate tourist trap.

There are places (private property) where collecting is permitted. Local inquiries can be made at the Monument or at rock shops in Kemmerer. There are several collectors who have special places where they get beautiful material. The Green River shales outcrop for many miles, so prospecting could be done with permission from local ranchers.

The northern half of the eastern edge of Wyoming to the west of the Black Hills uplift has some outcrops of the Jurassic dinosaur-bearing Morrison Formation, but is mainly marine and continental Cretaceous. Lance Creek is the site of a number of dinosaur discoveries, primarily *Triceratops*. The southeastern edge is mainly covered with Oligocene and early Miocene sediments. West of Douglas the ranchers will often let you prospect and collect on their land for a fee.

The Southern Rocky Mountain Region (AAPG Geological Highway Map Number 2) includes Utah, Colorado, Arizona, and New Mexico.

Northwestern Utah west of the Wasatch Front and south to Cedar City is part of the Basin and Range Province. Most of the area is covered by Quaternary lake deposits laid down on the bed of the Pleistocene Lake Bonneville which covered most of this area. The hills are Paleozoic marine rocks which are fossiliferous. Along the old Pleistocene shorelines are thick deposits of wave-built terrace deposits which contain a number of interesting fossils, including muskoxen.

To the south is the plateau area where the strata are quietly folded or severely faulted. Here the rocks range in age from late Paleozoic through Cretaceous with some early Tertiary remnants. Except for the marine sediments fossils are few and far between.

The northeastern part of Utah is dominated by the Uinta Mountains paralleling the Wyoming border. This is the only east-west–trending mountain range in the area. Here the folded beds have been eroded to expose a Precambrian core with upturned Paleozoic and Mesozoic sediments exposed on the sides.

South of the Uinta Mountains the Uinta Basin is filled with Paleocene and Eocene sediments. Further south, the southeastern corner of the state is covered with flat or gently folded Mesozoic rocks with patches of late Permian rocks showing through.

Western Colorado is essentially a continuation of what we've seen in eastern Utah, although there are extensive volcanics in the southwest. From the longitude of Rifle to the longitude of Denver the rocks are folded in many places and range in age from Precambrian intrusives and metamorphics to early Tertiary sediments and volcanics. In eastern Colorado the sediments are generally flat-lying and range in age from early Cretaceous to Pliocene. Weld and Logan Counties are rich in Oligocene land mammals.

Western and southern Arizona are made up of hills composed of Precambrian and younger metamorphics, Tertiary volcanics, and a few patches of Paleozoic and Mesozoic sediments. Surrounding these hills is a sea of Quaternary sediments that do contain Pleistocene fossils.

Separating this area from the plateau lands is a band of Tertiary volcanics and Precambrian metamorphics extending from Lake Meade to the New Mexico border. The northeastern moiety is the plateau country. The entire Paleozoic sequence of sediments is beautifully exposed in the Grand Canyon.

The massive cliff-forming Mississippian Redwall Limestone is the most fossiliferous formation exposed in the canyon. Capping the Grand Canyon and extending for miles is the Permian Kaibab Limestone which is also locally fossiliferous. Overlying the Kaibab Limestone is a thick Mesozoic series of formations. The red early Triassic Moenkopi Formation which forms the Painted Desert is locally fossiliferous, but the remains of large flatheaded amphibians—stegocenphalians—are difficult to collect and should be left for the expert. There are numerous fine-grained sandstone layers running through the red shales. The undersides of these sands often have raised ridges representing the negatives of mud cracks. Running across these drying mud surfaces stegocephalians and the strange reptile *Cheirotherium,* whose footprints resemble a human hand as there is a "thumb" on the outside of the foot, left their tracks.

The following article about the discovery of these trackways and other fossils, "The Hand Beast of Meteor Crater," is reprinted from the February 1974 issue of *Gems and Minerals Magazine:*

> It was shortly after sunup when we broke camp in the ponderosa pine forest west of Flagstaff, Arizona. The beginning of that early June day in 1938 promised to be hot and the journey a long one. By that night our five-man field party from the Museum of Paleontology at the University of California hoped to be at the summer digging grounds in central New Mexico.
>
> We breakfasted in Flagstaff and then turned east into the desert and the ovenlike air of an Arizona summer. Only a few miles out of town we saw the first signs advertising the wonders of Meteor Crater, a great scar on the earth's surface—more properly an astroblem (star blemish) resulting from the impact of a large meteor thousands of years ago. The two student members of our party had never been there before, so they set up a clamor to drive the few miles off the highway and see this "wonder of nature." With the skill of a long-time party leader, Dr. Charles Camp first suggested that there wasn't time and then bit by bit relented and gave the

children their way. At Rimmy Jim's beer bar, general store, and whatnot we turned south on the dirt road that led to the crater.

About a mile and a half before we reached the crater were a pair of small, low-lying buttes on either side of the road. Dr. Sam Welles, the Number Two faculty member, announced that he had visited this unmentionable thing so many censored times that he would probably unmentionable if he were forced to waste his censored time looking at this sickening sight again. Whereupon he bailed out of the truck and went off prospecting for fossils.

An hour or so later the faithful old GMC panelbody wheeled back down the dirt track off the crater rim, and we stopped to pick up the good doctor. Instead of looking like disaster about to strike, as he had when left at the roadside, he now closely resembled a cat who'd just polished off the last of the local canary farm. He had found fossil bone in the Moenkopi Formation! All thoughts of New Mexico vanished from our minds. This was very early Triassic bone from a formation that had never produced any before, and a whole new ball game was underway. Sam gets the credit for the discovery of the first bones from these barren red beds, but in our hearts we junior chums who had kicked our feet and screamed to see the crater knew the credit was really ours.

Camp was pitched, a run to Flagstaff produced permission from the Babbitt Brothers Land, Loans and Livestock Company to work on their land, and the dig began. Two weeks there produced a bonanza in the form of the skulls of some eighteen large, flat-headed amphibians new to science, and lots of skeletal parts. A rich deposit kept Berkeley field parties going back for many years until the site was dug out. But there was another fossil treasure hidden under the rocks. Yards of trackways of the mysterious early Triassic reptile *Cheirotherium,* the "hand beast," were found as the negative impressions along with negative impressions of mudcracks on the undersides of limy siltstone layers. When muddy surfaces dry out and crack, and then become sunbaked to an adobe brick hardness, the mud cracks and any footprints or other impressions made when the mud was soft may be preserved when a renewed source of water deposits fine, sandy sediments on the baked surface—before it melts. The sandy top layer hardens and preserves the features it covered up; on its underside, mud cracks look like raised lines, and footprints look like raised bumps.

One of the members of our party was Frank Peabody. He was a graduate student looking for a suitable problem for his doctoral dissertation. At Meteor Crator he found both his problem, and a project that was to occupy much of his scientific research time until his untimely death in 1958. At the quarry site and for miles around were limy sandstone layers a few inches thick bearing mud cracks and footprints on the undersides. The footprints included those of small reptiles, the flat-headed amphibians whose skulls we had found, and others, but mainly they were strange ones that looked like a human hand with a difference. When you look at a trackway the laterally extended thumb of each hand is on the outside instead of the inside! Frank had his problem; it was the true faith and there could be no other. He spent years working on these tracks and did a magnificent job of studying them.

The picture that emerged from his studies and others was something like this. Imagine a moderately stiff muddy surface left behind by a drying stream of muddy banks and shoals on the sides and in the midst of the channel. Imagine this river seeking the sea as it wandered across a broad, low coastal plain. At that time there was a shallow seaway just to the north, covering most of the area now included in Utah. This was during the early part of the Triassic Period, just at the beginning of the Age of Reptiles some two hundred thirty million years ago.

Above the Moenkopi Formation is the multicolored Chinle Formation. These sediments are notable for the petrified logs that are beautifully displayed at Petrified Forest National Park and for the long-nosed crocodilelike reptiles, the phytosaurs, which filled the crocodile niche in those days. Some skulls are over four feet long. Collecting is prohibited in the park, but there are ranchers in the area who will allow you to prospect on their land for a fee.

Above the Chinle Formation is a thick Mesozoic section. These beds have a few dinosaur remains, fossil leaves, and invertebrates in the marine sections. There are also scattered outcrops of Tertiary rocks which have produced very few fossils.

Northwestern New Mexico is underlaid by late Cretaceous rocks which produce some dinosaurs and the Paleocene and Eocene rocks of the San Juan Basin. Here there is abundant petrified wood and some mammal and reptilian remains.

The best collecting is in the late Paleozoic limestones that are exposed in a north–south band through the center of the state. I remember collecting walnut-sized spiny brachiopods which had been replaced by red silica from the limestones near Jemez Springs. Dissolving the limestone in acid left beautiful spiny specimens.

There is a wide band of late Tertiary deposits along the Texas border that should be extensively prospected.

The Northern Great Plains Region (AAPG Geological Highway Map Number 12) includes North Dakota, South Dakota, Nebraska, Minnesota, and Iowa.

North Dakota north and east of the Missouri River is largely covered with glacial deposits. Beneath these deposits there is a broad band of the early Paleocene Fort Union Formation extending from the Montana and Saskatchewan borders to just north of Bismark. The southern part of this band is the late Cretaceous Hell Creek Formation. Where exposed through the glacial deposits, land vertebrates could be found. East of these beds the area is covered with late Cretaceous sediments including the Pierre Shale and Fox Hills Sandstone. The Pierre Shale is marine while the Fox Hills Sandstone has both marine and continental facies. Both of these rock units are very fossiliferous in places. Where the marine Fox Hills outcrop one can collect beautiful ammonites by gathering nodules in the wheat fields after they have been plowed in the late summer.

South and west of the Missouri River the Fort Union Formation and the Cannonball Formation are exposed over most of the area. These could produce land mammals and reptiles as well as plants and fish.

The geology of South Dakota is dominated by the Black Hills on its western border. Here a Precambrian core of granitics and metamorphic rocks has been pushed up to form a dome mountain with a cover of sediments ranging in age from Cambrian to early Cretaceous. Fossils are not common in these strata, but invertebrates and fish have been collected.

Underlying virtually all of South Dakota is the Late Cretaceous Pierre Shale. These marine beds are very fossiliferous and among invertebrates are noted for beautiful ammonites including *Placenticeres, Scaphites, Baculites,* and *Sphenodiscus,* and the large clam *Inoceramus.* Fish and marine reptiles are also locally common. There are the giant marine lizards known as mosasaurs, the long-necked pleisiosaurs, and turtles up to 12 feet long.

"The Great Midwestern Ocean" reprinted from *Gems and Minerals Magizine* tells some things about collecting in these marine shales and limestones:

> Broad dry plains and high mountains are not the usual landscape conditions in the Midwest, Great Plains, and Rocky Mountain regions. During 90 percent of the last 600 million years of the Earth's history this area was mostly covered by shallow seas that overlay the core of the continent. The Midwest and the Great Plains are located on a fairly stable platform that has slowly risen and fallen like a giant elevator many times during the long history

of North America. The Rocky Mountain zone and the other mountainous areas to the west formed a broad trough which slowly sank for hundreds of millions of years to become the depository for thousands of feet of limestone, clay, silt, sand, and larger rock fragments. At the end of the Age of Reptiles, that is, at the end of the Cretaceous Period, these thick piles of sediments began to be folded into what were the beginnings of the present Cordillera or western mountain ranges.

The ancient seaway extended from the Arctic Ocean to the Gulf of Mexico (or their equivalents of the time). On the west an early uplift running from Idaho and Nevada southward into California and northward into Canada formed the western boundary of this sea and separated it from similar seas to the west. It was on the shores of this uplift that the herd of large ichthyosaurs was stranded which we described in "Here There Be Sea Monsters" (*G&M,* Oct. 1974). The shore lines fluctuated as the continent slowly pulsed up and down. At its greatest width the sea extended eastward to cover the Dakotas, Nebraska, a bit of southwestern Minnesota and western Iowa, western Kansas and Oklahoma, and out into the present Gulf of Mexico through Texas. To the north it was somewhat narrower in eastern Montana, Alberta, Saskatchewan, and the Northwest Territories.

As it happens with all things geological, they changed. The sea finally began to disappear at the end of the Cretaceous and only a small remnant persisted into the Paleocene, the first epoch of the Age of Mammals. Known as the Cannonball Sea for the large round concretions found in its sediments this last stand was made in parts of North and South Dakota.

The Great Midwestern Ocean was never very deep but it was filled with life. Many of the beds of mud, sand, and limestone deposited on its bottom are highly fossiliferous. It's the late Cretaceous sediments that hold the most interest, and particularly the Niobrara Chalk and the Pierre Shale. The Pierre Shale is named for Ft. Pierre on the west bank of the Missouri River in South Dakota. (The local pronunciation is "Peer.") These two formations are widely exposed today in western Kansas, Nebraska, northeastern Colorado, eastern Wyoming and the Dakotas.

In Kansas the Niobrara Chalk is particularly famous for the beautiful skeletons of marine reptiles, pterodactyls, ammonites, sharks, and the giant herring *Portheus* or *Xiphactinus*. Although not as well-publicized in the literature the Pierre Shale has much the same fauna in its western exposures where the Niobrara Chalk is thin and less fossiliferous. Most hobbyists are familiar with the beautiful specimens of the coiled ammonites *Scaphites* and *Placenticeras* and the straight ammonite *Baculites* from the Pierre Shale in South Dakota. Many of these specimens were found near the town of Wasta.

The marine life of the late Cretaceous must have been a spectacle, mostly of giant proportions. Among the reptiles were the long-necked plesiosaurs that sometimes reached a length of fifty feet. Their barrel-shaped bodies with the long neck and shorter tail perhaps resembled a turtle with a snake strung through it. For swimming there were four large paddles which slowly moved back and forth as the plesiosaur majesticially rowed its way through the sea. The long flexible neck ending in a small head with sharp conical teeth would dart about catching fish. We often think of a plesiosaur as resembling two elderly gentlemen slowly rowing a boat while a third stood in the bow casting for fish with a fly rod. The plesiosaurs were food gulpers, so to aid in digesting their fish feasts they swallowed pebbles to act as mill stones in their stomachs. Several skeletons have been found with a bucket's worth of acid-etched pebbles in the body cavity. Sometimes the source of these pebbles was hundreds of miles from the place where the skeleton was found. Mounted skeletons of plesiosaurs from these formations may be seen at the University of Kansas and the University of Nebraska museums, the South Dakota School of Mines Museum of Geology, and the Denver Natural History Museum. The University of California Museum of Paleontology, the California Academy of Science, and the Los Angeles County Museum of Natural History have skeletons of plesiosaurs from West Coast equivalents of the Great Midwestern Ocean.

The most ferocious marine reptiles were the mosasaurs, varanoid lizards who abandoned the land to live in the sea during the late Cretaceous. Some of these reached a length of thirty feet and had skulls over three feet long. The tails were high and narrow, effective sculls for swimming, while the legs became short paddles for balancing and steering. There is a handsome mounted skeleton of a Pierre Shale mosasaur in the School of Mines museum.

The late Cretaceous was also a day of triumph for the last gigantic sea turtles. *Archelon*, from the Pierre Shale near Rapid City, South Dakota, was a king-size leatherback about twelve feet long. The skeleton of this seagoing giant is displayed in the Yale University museum. It is missing one hind leg which shows a healed stump. Who knows whether it was a mosasaur or a large shark that missed its meal of turtle steak?

While these reptiles were ruling the seaways, there were pterodactyls overhead with wing spreads of thirty to thirty-five feet. These successful gliders ranged in time from the Jurassic through the Cretaceous when they finally gave way to the better organized birds. The pterodactyl wing was a membrane stretched between the greatly elongated fourth finger and the hind legs. With their hollow bones they didn't weigh very much, perhaps little more than ten pounds for an individual with a twenty-five-foot wing spread. These late Cretaceous species were toothless and must have taken off by spreading their wings to be lifted into the air by the wind. We believe that the large head crests sported by some late Cretaceous pterodactyls may have served as rudders to help keep them headed into the wind as they rode the waves to be lifted into the air. Their life mode suggests that they were warm blooded and probably supplied some sort of nurture for their young. These gliders of the Midwestern Ocean were until recently considered to be the largest of their kind. The title now goes to a non-fish eater found in Texas with an estimated wing span of fifty-one feet.

Perhaps the most interesting vertebrate in these beds is the flightless, wingless bird, *Hesperornis*. This six foot long bird lived like a modern seal, following schools of fish which it caught with its long toothed beak. It swam with powerful legs which, a bit like a loon's, were attached too far back on the body to be useful for standing on dry land.

While many sorts of fish and sharks are abundantly preserved in these beds, it is the super herring, *Portheus,* often attaining a length of twelve feet who is the piscatorial star. This fish is a prize catch on any fossil hunting expedition. One specimen was found in the Niobrara Chalk with another fish in its tummy large enough to make even the greediest fisherman happy.

In spite of these large vertebrates, the fossils most often sought by hobbyists are the beautiful ammonities and nautiloids from the shale formations. There are also clams, oysters, sea urchins, and even crabs to be found in these formations.

In the Great Plains there are literally hundreds of square miles of exposures where fossil prospecting may be rewarding. Generally speaking, on public land the collecting of invertebrates is allowed without permit although local inquiry is suggested. It is only vertebrates that are a no-no. Conscientious collectors who find good vertebrate material, while working these beds, should cover the exposed bones, mark the spot, and report the find to the nearest university or museum.

When properly approached many ranchers will allow people to

prospect in their pastures. You must remember that gates are to be left the way you found them, cattle are not to be molested, and that fire could mean the loss of a season's income for a ranch family.

If you are planning a collecting trip into the bottom of the Great Midwestern Ocean remember that the Spring and early Summer are wet, the mid- and late Summer is hot, but that the Fall is a beautiful time of year to look for fossils.

East of the Missouri River, which in the southern two-thirds of East River approximates the farthest southwest advance of the Pleistocene ice sheets, the terrain is thoroughly covered with glacial deposits. In the northern third of the East River the glacial deposits extend up to 50 miles west of the Missouri River.

Above the Pierre Shale in the west are broad exposures of the late Cretaceous Fox Hills Formation. This has both a continental and a marine facies. The marine facies is very fossiliferous in spots. Nodule collecting in plowed wheat fields produces beautifully preserved ammonites.

The Hell Creek Formation is widespread in the northwestern counties of South Dakota. This is the latest Cretaceous formation in the area. It is the formation which has produced so many spectacular dinosaurs in Montana. In South Dakota duckbills, *Triceratops,* bone heads, and *Tyrannosaurus* remains have been found.

In the northwest below the North Dakota border the late Cretaceous sediments are overlaid with early Tertiary strata which are locally very fossiliferous.

The world-famous White River Badlands or Big Badlands are exposed on both sides of the White River to the southeast of the Black Hills. The Oligocene mammalian fauna from these beds is probably the best on record. Heretofore unknown species are found every year. Much of this is part of Badlands National Park, Buffalo Gap National Grasslands, or the Pine Ridge Indian Reservation. Here collecting is prohibited without a permit, but there is private land where permission may be granted to collectors.

Minnesota is mostly covered with glacial deposits where Ice Age fossils may be found in lake and swamp deposits. Below the glacial sediments and exposed along river and stream courses are late Cretaceous marine deposits in the west. These are a combination of the sequence we found in South Dakota. Across the northern border, along Lake Superior and running in a wide northeast–southwest band from the Upper Peninsula to the middle of the state, are Precambrian rocks. Most of these are volcanics or metamorphics, but some do contain microscopic one-cell protists or moneras. Around the Twin Cities and extending southeast to the borders of Iowa and Wisconsin is a band of early Paleozoic rocks, Cambrian, Ordovician, and Devonian in age, which are often highly fossiliferous. Outcrops are found along the course of the Mississippi River and its tributaries.

Nebraska in many ways is like South Dakota, only more so. Underlying the entire state, except for a small area of Pennsylvanian rocks in the extreme southeastern corner, are Cretaceous marine and continental sediments. Overlying the Cretaceous beds in most of the state is a sequence of middle and late Tertiary beds which are often very fossiliferous. The eastern margin is covered with glacial deposits on the Cretaceous strata while the north-central area has extensive Quaternary sand dune deposits known as the Sand Hills.

In the west along the drainages of the White and the North Platte Rivers are outcrops of the Oligocene White River sequence and overlying Oligocene rocks. These are just as fossiliferous as their equivalents in South Dakota. Above these are Miocene sediments that in some places are virtual bone beds. Agate, Nebraska in the panhandle is the site of Agate Fossil Beds National Monument. Here, during the earliest Miocene, scores, if not hundreds, of rhinoceroses and other mammals perished in what may have been a series of dry years. Recently, in the northeastern part of the state several score rhinos, horses, and other mammals and birds were discovered buried in volcanic ash. These articulated skeletons were the victims of smothering in a great ash fall.

The Pleistocene deposits in the Sand Hills, in glacial deposits, and along river terraces have also produced an abundance of fossil remains.

Iowa is pretty much a repeat of Minnesota. Glacial drift covers much of the state and produces vertebrate fossils from lake and swamp deposits. Underlying these sediments are Paleozoic and Cretaceous beds that do carry fossils. The northwestern quadrant has a continuation of the late Cretaceous marine deposits seen to the north and west. Here is the eastern shore of the Great Midwestern Ocean. All of the other sediments are Paleozoic in age. To the northeast and east are Ordovician, Siluvian, and Devonian strata which are productive along the Mississippi River and its tributaries. The remainder of the state is covered with Mississippian marine limestone and Pennsylvanian coal swamp deposits. Both of these make excellent collecting areas.

The Mid-Continent Region (AAPG Geological Map Number 1) includes Kansas, Missouri, Oklahoma, and Arkansas.

Kansas is underlain by essentially flat-lying beds ranging in age from Pennsylvanian in the east to late Miocene in the west. Pleistocene deposits are scattered throughout the state, and there are glacial deposits in the east. The eastern quarter is made up of both marine and continental Pennsylvanian deposits. Invertebrates are common in the marine section, while plants may be found associated with the continental coal deposits. Further west the overlying Permian deposits form a narrow north–south strip north of Wichita and then spread westward to Meade County.

The north-central region has broad exposures of Cretaceous marine sediments. These, in part, are the famous Niobrara Chalk which produces plesiosaurs, marine turtles, and pterodactyls. There are also many invertebrates found here, including swarms of the free-swimming crimoid *Uintacrinus*. The western quarter is covered with late Tertiary and Quaternary deposits which are locally rich in mammals.

Missouri contains rocks ranging in age from Precambrian to Eocene. The center of this is in the Ozark Uplift south of St. Louis where the Precambrian is exposed. Cambrian and Ordovician rocks surround the core of the uplift. Beyond these, to the southwest, are Mississippian marine deposits which are very fossiliferous. To the north and far western edge of the state are marine and continental Pennsylvanian sediments. Plants may be found in association with the coal deposits. North of the Missouri River, glacial deposits cover most of the underlying formations.

During the early Tertiary the Mississippi Embayment extended as far north as the site of Cairo, Illinois. In the southeastern corner of Missouri, patches of the sediments deposited in this embayment show throughout the Quaternary deposits that cover the Mississippi River's valley.

Oklahoma is a combination of folding and flat. In the northeast corner, the west end of the Ozark Uplift, there are scattered outcrops of Ordovician rocks in the windows in rocks that are lumped together in the AAPG maps as Ordovician through Devonian. In the southeast, Oklahoma shares the Ouachita Uplift with Arkansas. Here the folded beds range from Devonian through Permian in age. Other folds and faulting continue to the west along the southern border. Most of the state is underlain by flat-lying or gently folded rocks. Along the Texas border, in the eastern half, are fossiliferous Cretaceous rocks.

North and west of the uplifts and dominating the eastern half of the state are Pennsylvanian rocks, some coal-bearing. The western half is underlain by Permian sediments. Some of these have produced reptilian remains. The panhandle is mostly made up of continental Miocene deposits with some Mesozoic sediments sticking through.

There are scattered Pleistocene deposits along the major river drainages which are often fossiliferous.

The eastern and southern sides of Arkansas are the site of early Tertiary deposits of the Mississippi Embayment. These are generally hidden by the Quaternary deposits of the Mississippi River and its tributaries. The Tertiary and some Cretaceous deposits are generally exposed to the west of Pine Bluff. Across the northern border is a strip of early Paleozoic rocks, most of these Ordovician in age. Across the center of the state is the Arkoma Basin filled with Pennsylvanian rocks. South of the basin is the Ouachita Uplift where sediments ranging in age from Ordovician to Pennsylvanian are exposed.

Fossils may be found in all of these sediments, particularly in the Mississippian limestones.

Because of its size and diversity, Texas rates AAPG Geological Highway Map Number 7 all to itself. Virtually the entire geologic column from late Precambrian through Pleistocene is represented and fossils are very abundant. I can remember one day in 1942, while attending a Tank Destroyer class at Camp Hood, Texas, filling my mussette bag with Cretaceous ammonites. I was sitting on a hillside watching a Piper Cub spotting plane demonstrating the adjustment of artillcry fire, but the fossils were more interesting.

Texas can be divided into five regions: the High Plains, covering the panhandle; north-central Texas and the Llano Area; east Texas and the upper Gulf Coast; west Texas; and southwest Texas and the lower Gulf Coast.

The western part of the High Plains region is also known as the Llano Estacado or the Staked Plains and is covered with late Miocene continental sediments that are locally fossiliferous. There is a large exposure of Triassic rocks to the north and west of Amarillo and late Permian to the east along the headwaters of the Canadian River. The Caprock Escarpment marks the eastern edge of the Staked Plains. Below this the eastern part of this region is underlain by Permian and Triassic sediments.

North-central Texas and the Llano Area has north–south exposures of Pennsylvanian and Permian rocks covering most of the region, with Cretaceous marine sediments to the east and south. In the southeastern corner is the Llano Uplift with exposures of Precambrian granites and scattered Paleozoics.

East Texas and the upper Gulf Coast has a band of Cretaceous sediments running along the Oklahoma border and southward along the west side to below Austin. To the east and south are bands of Cenozoic sediments ranging in age from Paleocene to Quaternary. These are interfingering sandstones and shales which are younger as you approach the Gulf Coast.

In west Texas the fossil-bearing rocks range in age from Cambrian to Pleistocene. The Big Bend area has been severely faulted, but important early Tertiary mammal fossils have been found. Any discoveries of these should be reported to the vertebrate paleontologists of the University of Texas in Austin.

Southwest Texas and the lower Gulf Coast area has fairly flat-lying Cretaceous sediments in the west. From San Antonio to the Gulf of Mexico is the continuation of the bands of Cenozoic sediments that we find on the upper Gulf Coast.

The AAPG Geological Highway Map has suggestions on where to look for fossils. The Permian limestones of the Guadalupe Mountains and the flat-lying Cretaceous marine deposits are the most productive beds in Texas.

The Great Lakes Region (AAPG Geological Highway Map Number 11) includes Wisconsin, Michigan, Illinois, Indiana, and Ohio. These states are underlain by fossiliferous Paleozoic deposits, making the area one of the best collecting areas in the world. Much of the region is covered by glacial moraines. These cover the older sediments but do contain numerous fossils of Pleistocene mammals.

Wisconsin is rather thoroughly covered by Pleistocene glacial deposits, but there are extensive Paleozoic outcroppings in the Driftless Area in the southwest and scattered about the state. The tenacious hunter might do well to search the unmetamorphosed later Precambrian rocks to the south of Lake Superior. In 1947 a rich late Precambrian metozoan fauna was discovered in the Ediacara Hills of southern Australia. Since its discovery, representatives of this fauna have been found throughout the world. A great contribution could be made by the fossil hunter who finds a similar fauna here in North America. Here and in the Belt series in the Rocky Mountains would seem to be the most likely prospects.

In the western part of the state there is a broad band of late Cambrian sediments between Eau Claire and Madison that contains good trilobite faunas. The Ordovician is widely exposed in the southwest and in a band extending to the northeast to Marinette. To the east, highly fossiliferous Silurian sediments form a narrow band from south of Fond Du Lac to the tip of the Door Peninsula on the eastern shore of Green Bay.

Michigan is also broadly covered with glacial deposits, but in cuts, quarries, and river banks fossils may be found. A thick Paleozoic section is buried beneath these glacial deposits. The Upper Peninsula is pretty much a continuation of the rocks seen in eastern Wisconsin. Along the northern edge of the Lower Peninsula are excellent Devonian exposures with a rich fauna.

As the Lower Peninsula is a broad structural basin, the earlier Paleozoic rocks are only exposed around the edges. Underlying the glacial deposits, which yield good Pleistocene mammal remains, are mostly Pennsylvanian sediments and some unnamed Jurassic red beds.

Illinois is again widely covered with glacial deposits which do contain Pleistocene mammals. To the north the Ordovician and Silurian deposits found in Wisconsin are exposed. The most important fossils are those found in the Pennsylvanian coal measures. Of these, the Mazon Creek fauna is by far the most diverse and interesting.

Starting over a hundred years ago, collection made along Mazon Creek started a fossil rush that continues today. The fossils occur in ironstone concretions that weather out of the Francis Creek Shale of middle Pennsylvanian age. Today, with extensive strip mining, the nodules may be found on the spoil piles. The Peabody Coal Company's Pit Eleven in Kankakee and Will Counties is a popular collecting site.

Over 70 genera of plants have been reported from the land facies of these beds. From the land and marine facies over 300 species of invertebrates and vertebrates have been listed. Most of the major phyla of invertebrates are represented with worms, spiders, myriapods, and insects. Of these, the insects are the most important, with 139 species. Among the vertebrates there are 18 kinds of fish, 7 amphibians, and a reptile. In all, 307 species of animals have been identified in the Mazon Creek fauna.

Most of Indiana is covered by glacial deposits, but Paleozoic rocks are exposed in river valleys and deep cuts. South of Terre Haute there is a broad expanse of Paleozoic sediments exposed. The oldest rocks are in the east and become younger to the west. In the southeast are late Ordovician and Silurian rocks which contain a good fossil fauna. In the center are Mississippian sediments with excellent graptolite faunas. To the west central are Pennsylvanian outcrops. As in other glaciated states the Pleistocene deposits are good hunting grounds for mammoths, mastodons, and other Ice Age mammals.

Much of Ohio is covered with glacial deposits which are locally very fossiliferous. The Pleistocene beach terraces along Lake Erie are good hunting grounds, as is almost any stream bank or sand and gravel pit.

Paleozoic rocks range from Ordovician to Pennsylvanian in age and are broadly exposed in the eastern part of the state from Lake Erie to the Indiana border near Cincinnati. All of this section is richly fossiliferous, particularly the late Ordovician rocks in the southwest.

The Devonian fish fauna is world-famous. Beautiful examples can be seen in the Cleveland Natural History Museum. The giant among these is the armoured *Dinicthys,* which may have reached a length of 30 feet.

The Northeastern Region (AAPG Geological Highway Map Number 10) includes New York, Vermont, New Hampshire, Maine, Massachusetts, Connecticut, Rhode Island, Pennsylvania, and New Jersey.

All of the Northeastern Region down to northern Pennsylvania and New Jersey was glaciated. Long Island, Cape Cod, Nantucket, and Martha's Vineyard are terminal moraines left by the last three advances of the ice—the Illinoian, Kansan, and Nebraskan glaciations.

Collisions between Europe and North America during the middle and late Paleozoic thoroughly crumpled and folded the Paleozoic rocks of this region. Only western New York and Pennsylvania escaped this disruption. When the present Atlantic Ocean opened, parts of Europe were left behind to form eastern New England and southeastern Canada, and most of what is now Scotland and Ireland left North America to go with Europe.

New York is a treasure trove of Paleozoic fossils. It was here that the first serious work on Paleozoic invertebrates was begun. In 1859 James Hall, New York State Geologist, published on the Paleozoic faunas he had been working on. Since then this has been a classic collecting area for Paleozoic fossils. Almost any outcrop may be examined, be it a highway or railway cut, a stream bank or quarry. Visits to the State Museum in Albany and the American Museum of Natural History in New York City are a must for fossil hunters.

The late Devonian is of particular interest, as it is from these beds that the famous forest at Gilboa was found and in the marine facies the beautifully preserved skeletons of the siliceous sponge *Hydnoceras* can be collected.

The Pleistocene should not be neglected as the remains of mammoths and mastodons are fairly abundant. Gravel pits and other excavations should be monitored.

Vermont has been badly folded, intruded by hot magmas, and metamorphosed. Early Paleozoic fossils ranging from Cambrian to Silurian in age can be found in those quartzites and marbles. Trilobites are locally abundant.

New Hampshire is not very fossiliferous because of the widespread occurrence of granitic and metamorphic rocks. Sediments were deposited here between late Precambrian and early Devonian times. Fossils are here, but they are not abundant.

Maine was thoroughly glaciated during the Pleistocene. Paleozoic sediments range from Cambrian to the Devonian. Generally these beds are not fossiliferous, but Ordovician shales and Silurian sediments have yielded a few fossils.

When one moves into Massachusetts the situation improves somewhat with sediments ranging from Precambrian to Cretaceous in age. Paleozoic fossils are sparse, with the exception of some locally fossiliferous Cambrian deposits. The Connecticut River Valley with its Triassic continental deposits extends into Massachusetts. In some areas there are abundant dinosaur and other reptile tracks, some dinosaur bones, and many fish and plants.

Although Connecticut has rocks extending in age from Precambrian to Jurassic, the only important ones are the Triassic continental deposits of the Connecticut River Valley. Here, cuts and quarries may be searched for dinosaurs, other reptiles, fish, and plants. Footprints and trackways of dinosaurs and other reptiles are common, but bones are very scarce.

Rhode Island offers little to the fossil hunter as it is underlain by igneous and metamorphic rocks. The fossil hunter would do well to concentrate on the Pleistocene glacial deposits and search for mammoths and mastodons.

Pennsylvania's geology is dominated by extensive deposits of Pennsylvanian sediments. Those laid down in coastal or inland swamps formed the coal measures which often yield coalified trunks and limbs of trees like the giant horsetail *Calamites;* the scale trees *Lepidodendron* and *Sigillaria;* and the tree ferns. This was a period of cyclic advances and retreats of the sea, probably caused by the interaction of southern hemisphere glaciation which removed water from the sea and surges of sea floor spreading which created new hot, expanded crust restricting the capacity of the oceanic basins.

When the land was flooded, marine limestones were deposited above the swamp deposits which were to be slowly converted to coal. These limestones may yield fossils where exposed in cuts and quarries.

Actually, there is a good Paleozoic section underlying Pennsylvania as well as some of the Triassic Newark series which contain plants.

As Pennsylvania is a mountainous state in a moist climate, the combination of topography and vegetation may discourage many fossil hunters.

New Jersey is completely different from the other states in this section. In the northwestern corner there are Paleozoic rocks exposed which are locally fossiliferous. In the same area are Triassic sediments with well-preserved fish and marine invertebrates.

There is a belt of Cretaceous sediments exposed from Raritan Bay on the northeast to Salem on the southwest. Here, late Cretaceous marine mollusks may be collected. The remainder of New Jersey is covered with Tertiary and Quaternary sediments.

The Mid-Atlantic Region (AAPG Geological Highway Map Number 4) includes West Virginia, Maryland, Delaware, Kentucky, Virginia, Tennessee, North Carolina, and South Carolina. The region can be roughly divided into four areas. On the east is the broad coastal plain and Piedmont area.

The coastal plain is made of flat-lying or gently dipping Cretaceous and Cenozoic sediments that become younger from west to east. The Piedmont is an area of Paleozoic intrusives and Precambrian and Paleozoic sediments, many of which have been metamorphosed. Behind the coastal plain and Piedmont is the heavily folded and faulted Appalachian Basin and Appalachian Mountains area. Here are rocks ranging from Precambrian to Mississippian in age. To the west is a broad plateau of flat-lying Pennsylvanian sediments which are rich in coal deposits. There are some late Mississippian beds showing through in western Kentucky and in a northeast–southwest line from northeastern Kentucky to south-central Tennessee. The Cincinnati Arch, which began to form during the Ordovician, is represented by two domes, the Jessamine Dome near Frankfurt, Kentucky and the Nashville Dome in Tennessee. Here Ordovician, Silurian, and Devonian rocks show through the Pennsylvanian sediments. The broad tops of the domes are composed of Ordovician rocks, while narrow bands of Silurian and Devonian sediments form the borders. To the far west in Tennessee, the Mississippi Embayment is represented by late Cretaceous and Cenozoic deposits.

West Virginia has the folded rocks of valley and range province along its eastern border. There are good Devonian faunas to be found in these rocks. Most of the state is part of the Allegheny Plateau, underlain by essentially flat-lying Pennsylvanian rocks. Here, dump piles from coal mines, cuts, and quarries may be prospected for plants and amphibians. The scouring rush *Calamites* and the scale trees *Lepidodendron* and *Sigillaria* are common.

Maryland is best known for the fossils of the Miocene Calvert Formation. Exposed along the cliffs facing Chesapeake Bay, these sediments are famous for the marine mammal remains and the teeth of the great white shark *Carcharodon*. Between Baltimore and Annapolis is a strip of early Cretaceous sediments of the Potomac Group. Here a large fauna of dinosaurs, crocodiles, turtles, and fish have been discovered.

The remainder of Maryland is composed of the Piedmont Plateau and the folded and faulted Appalachian basin and mountain system. Here the Ordovician, Silurian, and Devonian rocks produce the best fossils. Cuts and quarries are the main source of these fossils.

Delaware has little to offer the fossil hunter. Most of the state is covered with Quaternary alluvium which could produce Pleistocene animals. There are scattered outcrops of Cretaceous and Tertiary beds which could be prospected, particularly along road, railways, and canal cuts.

Kentucky is divided between the Appalachian Plateau and the Interior Lowlands. It is mostly underlain by Mississippian and Pennsylvanian rocks except for the Ordovician, Silurian, and Devonian rocks exposed in the Jessamine Dome at the north end of the Cincinnati Arch and some Cretaceous and Cenozoic sediments in the extreme west where the Mississippi Embayment covered what is known as the Jackson Purchase or the Purchase Area.

Excellent fossils may be found in the rocks of the Cincinnati Arch. A particularly interesting area is on the Ohio River near Louisville where an Ordovician coral reef, the Falls of the Ohio, is cut by the river for nearly a mile. Nearly 200 species of coral have been found here.

The Mississippian limestones are widely quarried and are a source of marine invertebrates. The spoil piles from Pennsylvanian coal mines should be searched for plants and amphibians.

Virginia has sediments ranging from late Cretaceous to Pleistocene in age. In the east are widespread areas of the Miocene Chesapeake Group which is locally rich in shark teeth, marine mammals, and invertebrates. To the west is the piedmont composed of intrusives and Precambrian rocks with scattered deposits of Triassic sediments of the Newark Group. Reptile tracks and dinosaurs are always worth looking for in these beds.

The folded Paleozoic rocks of the Appalachians are locally fossiliferous, and, as usual, cuts and quarries should be examined.

In the far west are flat-lying Pennsylvanian sediments which may be prospected for the usual coal measure plants and animals.

Landlocked Tennessee begins with the folded Paleozoic rocks of the Appalachians on the east. To the west is a broad expanse of Mississippian and Pennsylvanian sediments. This area is interrupted in the center by the Nashville Dome, the southern end of the Cincinnati Arch. Here Ordovician through Devonian rocks are exposed. The Ordovician and Silurian beds are particularly fossiliferous.

In the far west are the Cretaceous and Cenozoic deposits of the Mississippi Embayment.

North Carolina is divided between the Appalachians, the piedmont, and the coastal plain. The best fossil hunting is on the coastal plain. Here sediments range from Late Cretaceous to Quaternary in age. Marine invertebrates are common in the Cretaceous, Miocene, and Pliocene sediments. Both invertebrates and land mammals may be found in Pleistocene deposits. The Miocene marine sediments should be searched for marine mammals and shark teeth.

South Carolina is similar to North Carolina. The widespread deposits of the coastal plain are not very fossiliferous, but plants and invertebrates have been found. There are Tertiary deposits ranging in age from Eocene to Miocene in age. These are often very fossiliferous with an abundance of marine invertebrates and shark teeth.

The Southeastern Region (AAPG Highway Geological Map Number 9) includes Louisiana, Mississippi, Alabama, Georgia, and Florida. With the exception of the southern tip of the piedmont, Appalachians, and plateau jutting down into Alabama and northern Georgia, the entire region is part of the Mississippi Embayment and the Gulf and Atlantic Coastal Plains. Going toward the Atlantic Ocean and the Gulf of Mexico from this peninsula of Paleozoic rocks, the sediments become progressively younger. This peninsula was first surrounded by early Cretaceous seas and coastal plains. Throughout most of the Cretaceous and Cenozoic the continental interior shed sediments into the Mississippi Embayment and the coastal plains areas. There are both swampy shoreline facies and marine facies in these sediments.

One caution to the collector: Older fossils are sometimes reworked and deposited in younger sediments. This, of course, delights the creationists! Some years ago I was given a big collection of fossils from the Cretaceous Selma Chalk in Alabama. I used this collection for an identification and dating exercise for my historical geology and paleontology classes. Mixed into this typical Cretaceous fauna were a number of specimens of the blastoid *Pentremites,* a genus restricted to the Mississippian and Pennsylvanian. These nutlike echinoderms are tough and readily transported by streams without too much damage. Redeposited in the Selma Chalk, they had superficially taken on the milieu of a Selma fossil.

A great deal of Louisiana is covered by the Quaternary sediments of the Mississippi River and the Red River. In some of the northern counties there are exposed patches of late Cretaceous sediments peeking out as an island surrounded by a sea of Tertiary sediments. All of Louisiana was part of the Mississippi Embayment or the Gulf of Mexico during the late Cretaceous and all of the Cenozoic.

Almost all of Louisiana where there are outcrops of Tertiary rocks may be searched for fossils. Leaves and logs may be found in the coast plain facies, and in the marine facies are many rich invertebrate faunas. Perhaps the most exciting finds are the remains of the primitive whale *Zeuglodon* from the late Eocene deposits. I have been told that in some Gulf Coast areas, particularly in Alabama, stone walls were once built with the vertebrae of this great whale.

Except for a very small area in the extreme northeast, Mississippi was part of the Mississippi Embayment and the Gulf Coastal Plain. Mississippian beds are found in Tishomingo County which are fossiliferous. All the rest is Cretaceous or Cenozoic. The Mississippian period was not named for the state, but for a system of rocks described in the upper Mississippi Valley.

The northeastern corner of the state is underlain by late Cretaceous sediments with a rich invertebrate fauna. Moving westward and southward, each epoch of the Cenozoic is successively represented. The Tertiary formations represent both coastal plain and shallow marine facies. Invertebrates are common, while vertebrates are less common. The large vertebrae, teeth, and skull parts of the primitive whale *Zeuglodon* are common in the late Eocene Yazoo Formation. The late Paleocene–early Eocene Wilcox Group contains excellent leaf imprints.

Around Vicksburg the Pleistocene loess deposits contain a late Pleistocene Rancho la Brea–type fauna. Loess deposits are made of windborn dust. Characteristically it is yellowish in color, made up of unweathered rock flour, and forms vertical cliffs. In this case the loess is made of the glacial flour from the outwash plains of the retreating glaciers. It is easily dug and stands without timbering. At Vicksburg it was effectively used for entrenchments by the defending Confederates.

The northeastern quarter of Alabama is dominated by the southern end of the Appalachians. Pushing south through the northern border is the southern end of the Cincinnati Arch with exposures of rocks ranging from Cambrian through Devonian. Below this is a band of Mississippian rocks and then a triangle of the Pennsylvanian Pottsville Group extending south to Tuscaloosa. Arcing around this are a broad band of Cretaceous sediments and the Cenozoic deposits of the Mississippi Embayment and the Gulf Coastal Plain.

Alabama may be the most fossiliferous state in the Union, with fossil-bearing beds covering the Paleozoic through the Pennsylvanian, the upper Cretaceous, and all of the epochs of the Cenozoic. There is such a wealth of fossil material that one may prospect almost anywhere there is a natural or manmade outcrop.

Northern Georgia is underlain by the piedmont and the Southern Appalachians. This terrain extends south and east to Columbus, Macon, and Augusta. These cities are situated here as this is the fall line or the head of river navigation. Below this line is the late Cretaceous through Cenozoic sedimentary sequence.

The Appalachian folded area contains Paleozoic fossils, mainly Ordovician in age. The piedmont area is composed mainly of igneous and metamorphosed rocks, so it is virtually nonfossiliferous. The Cretaceous and Cenozoic rocks of the coastal plains are the primary source of fossils. The Cretaceous sequence contains marine invertebrates but should also produce vertebrate remains. The Cenozoic sequence has both marine and coastal plain facies. Both plant and land animal remains should be found in addition to the marine fauna.

The Florida panhandle and the northeastern part of the peninsula is underlain with the Paleocene through Quaternary sequence found in the other Gulf Coast states. Of this sequence, the late Eocene Ocala Limestone is particularly fossiliferous. In recent years a small Oligocene

White River–like mammalian fauna has been described from the peninsula. Excellent Miocene and "Pliocene" mammalian faunas have also been described in the Bone Valley Formation and other continental deposits. Every phosphate pit and limestone quarry should be searched for fossils. Mammoth and mastodon skeletons have been collected using scuba gear in a number of water-filled limestone sinks. Every bit of Florida which isn't under a lake or a swamp should be prospected.

Alaska and Hawaii are in the Circum-Pacific Edition of the AAPG Geological Highway Map Number 8.

Alaska is a vast country with rugged mountains and few roads. The Alaska Highway and the Trans-Alaska Pipeline have opened some areas to land travel, but most of Alaska is reachable only by boat or plane. Rocks ranging in age from Precambrian to Pleistocene are found throughout the state. Fossils may be found at almost any level, but the best are perhaps the Pleistocene vertebrates found in frozen soils and placer deposits.

The Pleistocene fauna is very Rancho la Brea–like with the addition of wolverine, muskox, caribou, moose, and Rocky Mountain goats. Some of these may be preserved with flesh. Years ago a skin of the head and a forefoot of a juvenile mammoth was recovered by the American Museum of Natural History.

Cretaceous amber weathering out of coal beds may contain insects.

Although Hawaii is our fastest-growing state, due to frequent lava flows adding to the island of Hawaii, it is probably the least fossiliferous as it is virtually all volcanic. There are some uplifted marine terraces on every island and fringing fossil reefs off of Oahu, Molokai, and Lanai. These contain Pleistocene invertebrates and should also produce fish, birds, and marine mammals.

Chapter Seven

WHERE TO GO FOR MORE INFORMATION

With the exception of Hawaii, each state has a geological survey or its equivalent. These public bureaus are the logical place to begin a search for more information. In addition to these sources the geology departments of local colleges or universities, natural history clubs, and local museums often have publications on geology and fossils.

Alabama

Geological Survey of Alabama
P.O. Drawer O
University, Alabama 35486

Alaska

Department of Natural Resources
Division of Mines and Geology
P.O. Box 5–300
College, Alaska 99701

Arizona

Arizona Bureau of Mines
The University of Arizona
Tucson, Arizona 85721

Arkansas

Arkansas Geological Commission
State Capitol
Little Rock, Arkansas 72201

California

California Department of Conservation
Division of Mines and Geology
Ferry Building
San Francisco, California 94111

Colorado

Colorado Geological Survey
220 Museum Building
Denver, Colorado 80202

Connecticut

The Connecticut Geological and National History Survey
Trinity College
Hartford, Connecticut 06101

Delaware

Delaware Geological Survey
University of Delaware
Newark, Delaware 19711

Florida

Florida Geological Survey
P.O. Box 631
Tallahassee, Florida 32304

Georgia

Department of Mines, Mining, and Geology
State Capitol Building
Atlanta, Georgia

Idaho

Idaho Bureau of Mines and Geology
Moscow, Idaho 83843

Illinois

Illinois State Geological Survey
Natural Resources Building
Urbana, Illinois 61801

Indiana

Indiana Department of Natural Resources

Iowa

Iowa Geological Survey
Geological Survey Building
Iowa City, Iowa 52240

Kansas

Kansas Geological Survey
The University of Kansas
Lawrence, Kansas 66045

Kentucky

Kentucky Geological Survey
Lexington, Kentucky 40507

Louisiana

Louisiana State Geological Survey
P.O. Box 8847
Baton Rouge, Louisiana 70803

Maine

Maine Geological Survey
Orono, Maine 04473

Maryland

Maryland Geological Survey
Latrobe Hall
Johns Hopkins University
Baltimore, Maryland 21218

Massachusetts

Geological Museum
Harvard University
Cambridge, Massachusetts 02138

Michigan

Geological Survey Division
Department of Conservation
Lansing, Michigan 48913

Minnesota

The Minnesota Geological Survey
The University of Minnesota
Minneapolis, Minnesota 55455

Mississippi

Mississippi Geological Survey
2525 Northwest Street
Jackson, Mississippi 39205

Missouri

Missouri Geological Survey
Rolla, Missouri 65401

Montana

Montana Bureau of Mines and Geology
Montana College of Mineral Science and Technology
Butte, Montana 59701

Nebraska

Conservation and Survey Division
113 Nebraska Hall
University of Nebraska
Lincoln, Nebraska 68508

Nevada

Nevada Bureau of Mines
University of Nevada
Reno, Nevada 89507

New Hampshire

Department of Resources and Economic Development
James Hall
University of New Hampshire
Durham, New Hampshire 03824

New Jersey

Department of Conservation and Economic Development
Bureau of Geology and Topography
520 East State Street
Trenton, New Jersey 08625

New Mexico

New Mexico State Bureau of Mines and Mineral Resources
Campus Station
Socorro, New Mexico 87801

New York

State Geologist
State Museum and Science Service
Albany, New York 12224

North Carolina

North Carolina Division of Mineral Resources
Raleigh, North Carolina 22607

North Dakota

State Geologist
The University of North Dakota
Grand Forks, North Dakota 58202

Ohio

Ohio Geological Survey
Columbus, Ohio 43216

Oklahoma

Oklahoma Geological Survey
The University of Oklahoma
Norman, Oklahoma 73069

Oregon

Oregon State Department of Geology and Mineral Industries
1069 State Office Building
Portland, Oregon 97201

Pennsylvania

Bureau of Topographic and Geological Survey
Harrisburg, Pennsylvania 17120

Rhode Island

Department of Natural Resources
Providence, Rhode Island 02903

South Carolina

South Carolina Division of Geology
P.O. Box 927
Columbia, South Carolina 29202

South Dakota

South Dakota Geological Survey Science Center
The University of South Dakota
Vermillion, South Dakota 57069

Tennessee

Tennessee Division of Geology
G-5 State Office Building
Nashville, Tennessee 37219

Texas

Bureau of Economic Geology
University of Texas
Austin, Texas 78712

Utah

Utah Geological and Mineralogical Survey
103 Geological Survey Building
Salt Lake City, Utah 84112

Utah State Paleontologist
Utah State Historical Society
307 West 2nd Street South
Suite 1000
Salt Lake City, Utah 84101

Vermont

Vermont Geological Survey
East Hall
University of Vermont
Burlington, Vermont 05401

Virginia

Virginia Division of Natural Resources
Charlottesville, Virginia 22903

Washington

Washington Division of Mines and Geology
Department of Natural Resources
Olympia, Washington 98504

West Virginia

West Virginia Geological and Economic Survey
P.O. Box 879
Morgantown, West Virginia 26505

Wisconsin

Wisconsin Geological and Natural History Survey
115 Science Hall
University of Wisconsin
Madison, Wisconsin 53141

Wyoming

Geological Survey of Wyoming
University of Wyoming
Laramie, Wyoming 82070

AFTERWORD

I hope this brief encounter with fossil collecting will open a wonderful new door for you. Please strive for the status of an amateur—a professional level of work that is much more rewarding than being a hobby collector.

Join your local mineral, rock, or fossil club. Volunteer to work at your local museum and become acquainted with the professionals in the area.

I will be glad to answer your questions or identify your discoveries.

Good luck and good hunting.

J. R. Macdonald
The Flying Aepinacodon Ranch
Harmony Heights
Rapid City, South Dakota 57701

Two old pros, Utah's James Madsen and the writer, discussing the day's success in a dinosaur quarry.

Index